Energy Science, Engineering and Technology

BIOFUELS FROM FISCHER-TROPSCH SYNTHESIS

ENERGY SCIENCE, ENGINEERING AND TECHNOLOGY

Air Conditioning Systems: Performance, Environment and Energy Factors
Tobias Hästesko and Otto Kiljunen (Editors)
2009. 978-1-60741-555-8

Oil Shale Developments
Ike S. Bussell (Editor)
2009. 978-1-60741-475-9

Power Systems Applications of Graph Theory
Jizhong Zhu (Author)
2009. 978-1-60741-364-6

Bioethanol: Production, Benefits and Economics
Jason B. Erbaum (Editor)
2009. 978-1-60741-697-5

Energy Costs, International Developments and New Directions
Leszek Kowalczyk and Jakub Piotrowski (Editors)
2009. 978-1-60741-700-2

Handbook of Exergy, Hydrogen Energy and Hydropower Research
Gaston Pélissier and Arthur Calvet (Editors)
2009. 978-1-60741-715-6

Wind Turbines: Types, Economics and Development
Gerard Martin and Jeremie Roux (Editors)
2009. 978-1-60741-849-8

Computational Techniques: The Multiphase CFD Approach to Fluidization and Green Energy Technologies (includes CD-ROM)
Dimitri Gidaspow and Veeraya Jiradilok (Authors)
2009. 978-1-60876-024-4

Cool Power: Natural Ventilation Systems in Historic Buildings
Carla Balocco and Giuseppe Grazzini (Authors)
2009. 978-1-60876-129-6

Radial-Bias-Combustion and Central-Fuel-Rich Swirl Pulverized Coal Burners for Wall-Fired Boilers
Zhengqi Li (Authors)
2009. 978-1-60876-455-6

Handbook of Sustainable Energy
W. H. Lee and V. G. Cho (Editors)
2010. 978-1-60876-263-7

Jatropha Curcas as a Premier Biofuel: Cost, Growing and Management
Claude Ponterio and Costanza Ferra (Editors)
2010. 978-1-60876-003-9

Photovoltaics: Developments, Applications and Impact
Hideki Tanaka and Kiyoshi Yamashita (Editors)
2010. 978-1-60876-022-0

A Solar Car Primer
Eric F. Thacher (Authors)
2010. 978-1-60876-161-6

Thin-Film Solar Cells
Abban Sahin and Hakim Kaya (Editors)
2010. 978-1-60741-818-4

Advanced Materials and Systems for Energy Conversion: Fundamentals and Applications
Yong X. Gan (Author)
2010. 978-1-60876-349-8

Nuclear Fuels: Manufacturing Processes, Forms and Safety
Antoine Lesage and Jérôme Tondreau (Editors)
2010. 978-1-60876-326-9

Ethanol Biofuel Production
Bratt P. Haas (Editor)
2010. 978-1-60876-086-2

Biodiesel Handling and Use Guide
Bryan D. O'Connery (Editor)
2010. 978-1-60876-138-8

Biodiesel Fuels Reexamined
Bryce A. Kohler (Editor)
2010. 978-1-60876-140-1

Syngas Generation from Hydrocarbons and Oxygenates with Structured Catalysts
Vladislav Sadykov, L. Bobrova, S. Pavlova, V. Simagina, L. Makarshin, V. Parmon, Julian R. H. Ross and Claude Mirodatos (Authors)
2010. 978-1-60876-323-8

Corn Straw and Biomass Blends: Combustion Characteristics and NO Formation
Zhengqi Li (Author)
2010. 978-1-60876-578-2

Introduction to Power Generation Technologies
Andreas Poullikkas (Author)
2010. 978-1-60876-472-3

CFD Modeling and Analysis of Different Novel Designs of Air-Breathing Pem Fuel Cells
Maher A.R. Sadiq Al-Baghdadi (Author)
2010. 978-1-60876-489-1

A Sociological Look at Biofuels: Understanding the Past/Prospects for the Future
Michael S. Carolan (Author)
2010. 978-1-60876-708-3

Bio Resource-Based Energy for Sustainable Societies
K.A. Vogt, D.J. Vogt, M. Shelton,
R. Cawston, L. Nackley et al.(Authors)
2010. 978-1-60876-803-5

Direct Methanol Fuel Cells
A. S. Arico, V. Baglio and V. Antonucci (Authors)
2010. 978-1-60876-865-3

Ejectors and their Usefulness in the Energy Savings
Latra Boumaraf, André Lallemand
and Philippe Haberschill (Authors)
2010. 978-1-61668-210-1

Utilisation and Development of Solar and Wind Resources
Abdeen Mustafa Omer (Author)
2010. 978-1-61668-238-5

Coalbed Natural Gas: Energy and Environment
K.J. Reddy (Editors)
2010. 978-1-61668-036-7

Thin Film Solar Cells: Current Status and Future Trends
Alessio Bosio and Alessandro Romeo (Editors)
2010. 978-1-61668-326-9

Sustainable Resilience of Energy Systems
Naim Hamdia Afgan (Author)
2010. 978-1-61668-483-9

Storage and Reliability of Electricity
Edward T. Glasby (Editor)
2010. 978-1-61668-248-4

Microalgae for Biofuel Production and CO_2 Sequestration
Christopher Lan and Bei Wang (Authors)
2010. 978-1-61668-151-7

Geothermal Energy: The Resource Under our Feet
Charles T. Malloy (Editor)
2010. 978-1-60741-502-2

Wind Power Market and Economic Trends
Martin W. Urley (Editor)
2010. 978-1-61668-243-9

Stages of Deployment of Syngas Cleaning Technologies
Filomena Pinto, Rui Neto André and I. Gulyurtlu (Authors)
2010. 978-1-61668-257-6

High Temperature Electrolysis in Large-Scale Hydrogen Production
Yu Bo and Xu Jingming (Authors)
2010. 978-1-61668-297-2

Biofuels from Fischer-Tropsch Synthesis
M. Ojeda and S. Rojas (Editors)
2010. 978-1-61668-366-5

Transient Diffusion in Nuclear Fuels Processes
Kal Renganathan Sharma (Author)
2010. 978-1-61668-369-6

Coal Combustion Research
Christopher T. Grace (Editor)
2010. 978-1-61668-423-5

Shale Gas Development
Katelyn M. Nash (Editor)
2010. 978-1-61668-545-4

Coal Combustion Research Advances
Juwei Zhang (Author)
2010. 978-1-61668-935-3

Energy Science, Engineering and Technology

BIOFUELS FROM FISCHER-TROPSCH SYNTHESIS

M. OJEDA
AND
S. ROJAS
EDITORS

Nova Science Publishers, Inc.
New York

For permission to use material from this book please contact us:
Telephone 631-231-7269; Fax 631-231-8175
Web Site: http://www.novapublishers.com

Library of Congress Cataloging-in-Publication Data

Biofuels from Fischer-Tropsch synthesis / editors, M. Ojeda and S. Rojas.
p. cm.
Includes index.
ISBN 978-1-61668-366-5 (softcover)
1. Biomass energy. 2. Fischer-Tropsch process. I. Ojeda, M. (Manuel), 1960- II. Rojas, S., 1962-
TP339.B5436 2009
662'.88--dc22

2010012149

Published by Nova Science Publishers, Inc. ✝ New York

CONTENTS

PREFACE

The world demand of light and middle distillate petroleum products (mainly gasoline and diesel) is increasing dramatically, especially within the transportation sector. Eventually, such requirement will surpass refineries capacity to supply fuels and hydrocarbon feedstock. This situation has encouraged the exploration of other sources of petroleum products rather than merely sticking to conventional oil sources. Within this scenario, natural gas, coal and biomass appear as suitable carbon sources to be converted into the highly demanded hydrocarbon feedstock through a demonstrated two-step technology: formation of synthesis gas (carbon source $\rightarrow H_2 + CO$) and Fischer-Tropsch synthesis (FTS), which consists in the CO hydrogenation reaction to form hydrocarbons ($H_2 + CO \rightarrow$ hydrocarbons). The FTS provides thus an alternative route for the production of clean transportation fuels and high molecular weight hydrocarbons from natural gas (gas-to-liquids, GTL), coal (coal-to-liquids, CTL) and/or biomass (biomass-to-liquids, BTL). This process takes place at high temperatures and moderate pressures in fixed bed or slurry reactors. Synthesis gas mixtures can be converted into useful fuels and petrochemicals with Fe- and Co-based catalysts. In particular, iron-based catalysts are interesting for this process because of their low cost, high activity, flexible product distribution, and possibility of using coal- and biomass-derived synthesis gas with low H_2/CO ratios. Moreover, Fe-based catalysts offer the possibility of using synthesis gas containing high amounts of CO_2. The catalytically active Fe phase has been the focus of intense research. It is mostly accepted that Fe carbides play an active role in the FTS, although the fact that iron suffers a reconstruction process from the oxide phase to a mixture of phases (metallic iron, iron oxides and iron carbides) under reaction conditions hinders the unequivocal identification of the real active sites. There is a lack of effective characterization techniques under typical reaction conditions in the FTS (1-4 MPa, 450-570 K). Therefore, most of the characterization studies have focused on catalytic precursors and spent catalysts. The situation is still more complicated considering that the presence of chemical promoters and supports affects the Fe phase and structure during the reaction. In this chapter, the FTS technology, namely reactor types, reaction conditions and typical catalysts, will be briefly reviewed. The core of this book is devoted to the description of active sites and reaction intermediates with Fe-based catalysts. Nevertheless, the most relevant aspects of Fischer-Tropsch Synthesis with Co and Ru catalysts are also considered.

Chapter 1

FUNDAMENTALS OF SYNGAS PRODUCTION AND FISCHER-TROPSCH SYNTHESIS

J.M. González-Carballo and J.L.G. Fierro
Department of Structure and Reactivity, Institute of Catalysis and Petrochemistry (CSIC); C/ Marie Curie 2, 28049 Madrid, Spain.

1.1. INTRODUCTION

The Fischer-Tropsch Synthesis (FTS) was born as the scientific response to the increasing demand of liquid fuels in Germany in the early 20th century. The production of synthetic fuels is a relatively complicated process with three different steps: (*i*) production of clean H_2/CO mixtures (synthesis gas or syngas) from coal, natural gas or biomass; (*ii*) the catalytic transformation of syngas into a hydrocarbon pool (FTS); (*iii*) upgrading of the hydrocarbon pool *via* a traditional refinery approach. In the early days of this process, coal was converted into liquid fuels (*coal-to-liquids*, CTL). Afterwards, natural gas was predominantly used as the reactants source (*gas-to-liquids,* GTL). Recently, the use of biomass has been demonstrated to be a promising route to obtain synthetic fuels (*biomass-to-liquids*, BTL). In this context, the BTL process is considered as a key catalytic approach for the development of the thermochemical platform in biorefineries aiming to produce the so-called second generation biofuels (fuels from lignocellulosic biomass) [1].

Carbon, natural gas and oil are the primary energy source in the modern society. The transportation sector accounts for *ca.* 35 % of the energy consumed (mainly as liquid fuels) in the World [2,3]. Furthermore, the demand in this sector is expected to increase during the next years. World use of liquid fuels will grow from 83.6 million barrels oil equivalent per day in 2005 to 95.6 million barrels per day in 2015 and 112.5 million barrels per day in 2030 [2]. Such demand will be mostly met by traditional oil sources. However, the finite reserves of crude oil and their uneven distribution results in volatile prices. Moreover, security supply issues, particularly for the western economies, and social conflicts could trigger oil-wars. Clearly, alternative sources of liquid fuels other than the traditional oil are needed to fulfill the energy requirements in the future. Furthermore, society appreciation and consciousness about the rational utilization of energy is changing dramatically. The average citizen is now aware of the importance of an efficient use of the available energy and the idea of *saving* energy is penetrating into our lifestyle. Arguably, the starting point of this energetic conscience is related to environmental issues, particularly CO_2 emissions and global warming effect. In this sense, the alternative liquid fuel source should be environmentally friendly. Finally, we must consider the utilization of the immense coal and strained natural gas reserves. Question arises as to whether it should be possible to use those reserves for energy purposes, particularly in the transportation sector, without a negative environmental impact. In other words, do we have any technology to obtain liquid hydrocarbons fuels from CO_2 neutral sources? The answer to these question leads again to the Fischer Tropsch Synthesis as a promising option for the supply of synthetic fuels.

The synthesis of methane from synthesis gas is known since the early 20th century when Sabatier and Senderens described the catalytic transformation of CO and CO_2 into methane and water [4]. This reaction was conducted in the presence of Ni or Co catalysts at 523 K and atmospheric pressure. Later, Fischer and Tropsch went further and described the production of hydrocarbons from syngas using alkali-iron catalysts operating at 600-700 K and 10-15 MPa [5]. The process was named as the *Synthol* process. These authors continued their investigations and obtained liquid hydrocarbons with Co-Fe and Cu-Fe catalysts operating at 400-450 K and atmospheric pressure to produce a pool of hydrocarbon gases and liquids. This process became known as the Fischer-Tropsch Synthesis and it is the key reaction involved in the transformation of coal, natural gas or biomass into hydrocarbons. This approach was rapidly implemented at the industrial level. Thus, IG Farben, Ruhrchemie and other chemical companies started the industrial production of synthetic liquid fuels in the mid 1930′s. By the time World War II ended, several FTS plants were already under operation and the annual production reached 4.1 million barrels in 1944. Thereafter, synthetic fuel production in Germany was prohibited, and FTS and coal hydrogenation plants were dismantled. Although Germany researchers and industry were originally the pioneers in the field of synthetic fuels, Sasol (South Africa) is the chief actor in the modern industry. The interest shown by South African companies in the FTS process started in the 1930′s. The first Sasol FTS plant, located in Sasolbourg, was operative in the mid 1950's with coal as raw material and Fe materials as catalysts. This plant started to produce chemicals instead of liquid fuels in the 1960′s. Some years later, two coal-based FTS plants were constructed in Secunda (South Africa) to produce more than 10 times the amount of fuel of the former plant. The higher production capacity was partially due to an improved reactor design, using circulating fluidized bed (CFB) reactors or fluidized bed reactors (SAS, Sasol Advanced Synthol). A further upgrade of the FTS process was the implementation of the slurry phase reactors in the 1980's. Sasol uses currently this type of reactor in Sasolburg with Fe catalysts to produce long chain hydrocarbons. This reactor technology has been also selected by Sasol in the plant to be constructed in Qatar that operates with natural gas as syngas source with cobalt catalysts. Other companies, such as ExxonMobil, Statoil, Syntroleum or ConoccoPhilips, use modified slurry reactors to produce synthetic fuels from syngas.

Although the first FTS plants used coal as synthesis gas source, natural gas is currently the preferred option because the hydrogen content is higher than that of coal-derived syngas. Moreover, large strained natural gas reserved could be available in market by transforming natural gas into liquid hydrocarbons. Nowadays, many studies have focused on the feasibility of FTS processes from biomass. This would be an ideal option as raw source since biomass is or could be widely available in many areas of our planet and more importantly, since biomass is renewable, CO_2 neutral emission fuels could be obtained. Choren (Germany) has recently opened a FTS plant with biomass as a syngas source, expecting to produce fuels from 2009 onwards [6]. The plant combines gasification of biomass (Carbo-V process) and the production of diesel and waxes (further cracked to maximize diesel production) by means of the FTS process (Choren labeled its FTS diesel as Sundiesel®). Other countries with vast biomass resources, such us Norway, Sweden or New Zealand, are currently studying the feasibility of building BTL plants with the gasification plus FTS technology.

1.2. Syngas Production, Cleaning, and Conditioning

1.2.1. Syngas Production

As mentioned above, syngas (H_2/CO mixtures in different ratios) can be obtained from several sources such as coal, natural gas or biomass (Figure 1.1). The main advantage of biomass as syngas source is the renewable character, yielding thus CO_2-netural fuels.

Biomass is a broad term that includes wood, wood wastes, agricultural crops and its by-products, municipal solid waste, animal wastes including food processing and algae. The chemical structure of biomass comprises carbon, hydrogen, oxygen and nitrogen. The components of biomass include mainly cellulose, hemicelluloses, lignin, lipids, proteins, sugars, starches, and water. Two main strategies are currently available to convert biomass into fuels. Thus, biomass can be subjected to biological or chemical treatments, such as hydrolysis and transesterification, producing sugars and biodiesel, respectively. Thermal treatment of biomass form a volatile fraction comprising gas and vapour, and a liquid residue, char and tar. Three biomass treatment processes are identified depending of the relative amount of oxidant used with respect to that of the biomass: *i*) *combustion*, when the process is conducted in excess of oxidant; *ii*) *pyrolysis*, when no oxidant is used, or its concentration is well below the equivalence ratio (ER), *i.e.*, the amount of oxidant added relative to the amount of oxidant required for stoichiometric combustion; and *iii*) *gasification*, when the process is conducted in a concentration of oxidant close to the equivalence ratio.

Pyrolysis, known since ancient Egyptian times, is the thermal decomposition of biomass in the absence of oxygen to produce solids (char), liquids (tar) and gaseous products [7]. The liquid and gas products can be used in engine and turbine devices for power applications. Typically, liquid production from pyrolysis of biomass is maximized between 620 and 770 K. The charcoal yield increases with decreasing temperature and residence time. Depending on the reaction conditions, the pyrolysis processes can be divided in conventional pyrolysis, fast pyrolysis, and flash pyrolysis. *Conventional pyrolysis* occurs at slow heating rates and yields solid, liquid, and gas products. *Fast pyrolysis* (also known as thermolysis) is aimed at maximizing the fraction of liquid products, amounting to 60-70 wt.% of the total pool. Typical reaction conditions includes residence time of 0.5-10 s, heating rate of

10-200 K s^{-1}, particle size below 1mm, and temperature range between 840 and 1250 K. By modifying the reaction conditions, in particular the temperature and heating rate, the gaseous fraction can be increased [7]. Finally, *flash pyrolysis* demands a short residence time (<0.5 s), a high heating rate (>1000 K s^{-1}), and a temperature range between 1050 and 1320 K.

In general, the liquid fraction obtained from biomass pyrolysis is referred to by different names, such as bio-oil, bio-crude oil, bio-fuel oil, wood oil, liquid smoke, or pyroligneous tar. This liquid phase is a dark brown viscous oil containing organic acids, aldehydes, ketones, phenols, anhydro-sugars, pyrolytic lignin, and other organic fragments along with water [8].

To the purpose of this book, the most interesting thermochemical route to transform biomass is gasification. It can be defined as the thermal treatment that produces a high amount of gaseous products (syngas) and a small quantity of char and ash. Gasification breaks down the raw material (biomass or coal) into its components usually by using water steam and controlled amounts of oxygen at high temperature and pressure, obtaining a gaseous fuel. The process is in fact a partial oxidation of a solid fuel. The origins of gasification can be traced back to 1812. Almost a million gasifiers were used during World War I to run vehicles using wood as a fuel [7]. Today, biomass gasification can be pictured as a process to produce liquid fuels and chemicals via synthesis gas.

Gasification is usually carried out at high temperatures in order to optimize the production of the gaseous fraction. The yield of biomass gasification comprises a mixture of H_2 and CO along with variable amounts of CO_2, H_2O, N_2 (in high amount when air is used as oxidant), CH_4, small hydrocarbons, tars, char, inorganic materials and ash. A number of gasifying agents are available, including air, steam, steam-oxygen, air-steam, oxygen-enriched air, etc. Although biomass gasification at large scale has been proved successful, its price is higher than traditional gasification of fossil fuels [9].

One of the most important problems derived from biomass gasification is how to deal with tar, which is a complex mixture of 1-ring to 5-ring aromatic compounds along with oxygen containing molecules and polycyclic aromatic hydrocarbons.

biomass + oxidant (H_2O, O_2, O_2/N_2) → CO, H_2, CO_2, CH_4 + hydrocarbons (tar, char and ash) + H_2S + HCN + NH_3 + HCl + inorganic materials (Si, Al, Fe, Ti, Ca, Mg)

In the case of biomass, several steps should precede gasification, such as drying and size reduction. Elimination of sulfur containing compounds is necessary in all cases. Purification and adjustment implies removal of further sulfur compounds, nitrogen containing molecules (nitrogen oxides, HCN or ammonia), unreacted or produced methane (from methane reforming or coal gasification respectively), CO_2, N_2 (when air is used during gasification), tar, etc. This leads to the production of syngas, a mixture mainly consisting of CO and H_2. The desired ratio H_2/CO should be tuned by subjecting the mixture to a water-gas shift (WGS) process.

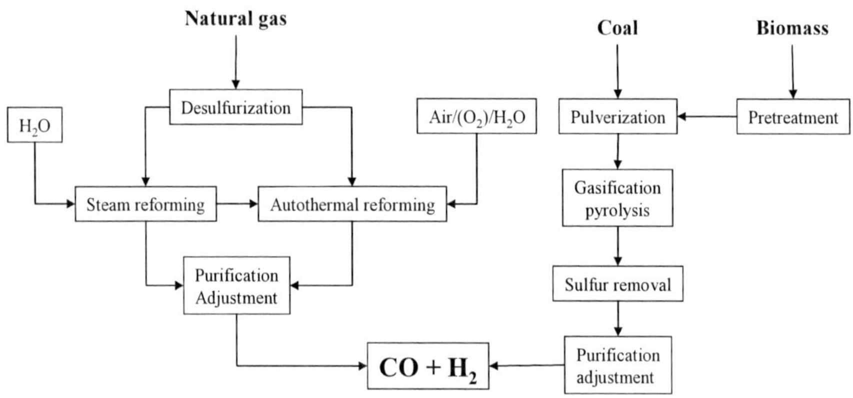

Figure 1.1. Steps in syngas production from natural gas, coal and biomass.

The synthesis gas can be produced from coal and oxygen (with added water) at high temperatures. The process can be divided in the following three steps: (*i*) thermal decomposition of coal or pyrolysis to form synthesis gas with variable amounts of CO_2, CH_4, H_2S, NH_3 and/or N_2; (*ii*) combustion of the impurities; and (*iii*) gasification with H_2O and CO_2 to produce the H_2/CO mixtures.

An extensive review of gasifiers manufacturers in Europe, USA and Canada can be found elsewhere [9]. According to Balat *et al.* [7], Fixed Bed Gasifiers (FBG) are the most suitable for biomass gasification. Biomass moves as a plug in the reactor and the oxidant agent is in close contact to the biomass particles. Fixed bed gasifiers are fed from the top side in either updraft or downdraft configurations. In the Atmospheric Updraft Gasifiers (AUG), the air or oxygen passes in a direction counter current to the flow of biomass. They can be easily scaled up but the amount of tar is very high limiting their market in power applications since a severe gas cleaning is required due to the presence of tar. On the other hand, they are easy to scale up and they can find application in biomass gasification for syngas synthesis. Atmospheric Downdraft Gasifiers (ADG) have found wide applications for small scale power plants. Tar issues are as important as with AUG.

Fluidized bed gasifiers take advantage of the excellent mixing characteristics and high reaction rates of this method of gas-solid contact. Typically, they operate at 800-1000°C, a temperature range limited by the melting properties of the bed material. Therefore, they are not suitable for coal gasification.

Atmospheric Circulating fluidized bed gasifiers (ACFBG) are the appropriate option for large-scale syngas production. These reactors are particularly suitable for the gasification of coal or biomass. They have proven reliable with a variety of feedstock and are easy to scale up to 100 MWth. It appears to be the preferred option for large scale applications, being used by companies such as TPS, Foster Wheeler or Lurgi.

Atmospheric Bubbling Fluidized Bed Gasifiers (ABFBG) is a reliable option with a variety of feedstock at pilot scale. They have found commercial application in medium scale (25 MWth). Pressurised Fluidized Bed Gasifiers can be circulating (PCFBG) or bubbling (PBFBG) are complex in operation and are more expensive due to the demand of pressurised vessels. They operate with compressed fuel, and consequently, an auxiliary pressurising unit is in order. Nevertheless, the obtained gas is pressurized and can be fed to the Fischer-Tropsch or to the combustion chamber of the gas turbine unit.

Gasification should be conducted in the presence of oxygen. If air is used instead, the process cost is lower, but the syngas is enriched in N_2, which should be removed before reaching the FT reactor. In contrast, very large and costly plants are needed for the separation of oxygen from air if the gasification step is conducted with pure oxygen. The most adequate option is through cryogenic plants, although the cost of the overall process increases significantly. An interesting option is the use of pressurized gasifiers because the size of the reactors can be thus decreased considerably. Pressurized gasifiers can have a significant impact in large scale syngas production; the gasifier may be smaller per throughput, and therefore, a larger maximum capacity is possible. In addition, no additional compression is required because syngas is already obtained at high pressure.

The gasification process to obtain biomass (or hydrogen) is usually conducted in the presence of a catalyst to decrease the operation temperature. Usually, solid catalysts consisting in a metal deposited onto a high area inorganic carrier are employed. Certain promoting agents are usually added as minority components to increase the activity and/or stability of the catalyst. Most catalytic formulations employ Ni as the active phase; however, studies with Au, Co, Fe, Cu, Ir, Pd, Pt, Zn, Ru, and Rh are also available. Among those, Rh, Ni and Pt are proposed as the most active metals. Needleless to mention, cost issues related to the use of noble metals make Ni as the most feasible option. Support materials provide mechanical strength and increase metal dispersion, which reverts in a superior utilization of the metal phase per gram. This is particularly relevant for the use of noble metal catalysts.

Tar removal still remains the main technical challenge for the successful commercialization of biomass gasification technologies. Complete tar elimination is concentrated in three main approaches: scrubbing, catalytic reforming (followed by scrubbing), and hot gas clean up. Scrubbing is ruled out as a viable option because it is only effective during short periods and, more importantly, it could create a serious environmental issue because of the large amounts of condensate product. Catalytic conditioning (tar cracking) is the preferred option. [10] It is based on the use of calcined dolomites ($CaCo_3$-$MgCo_3$) or of steam reforming based catalysts, typically based on Ni supported on inorganic carriers. Dolomite has a poor attrition resistance so other materials such as olivine (a silicate containing Fe and Mg) is also used. Neither of them displays reasonable performance in methane reforming so they are modified by incorporating Ni by impregnation.

Synthesis gas can be also produced from natural gas, which is a complex mixture containing different hydrocarbons (mainly methane), although other gases (N_2, CO_2, H_2O, H_2S, He, etc.) are also present. Therefore, natural gas must be purified before it can be converted into liquid fuels. The main components that need to be removed are those containing sulphur because these can poison the catalysts. Synthesis gas can be obtained from methane *via* the following processes [11-13]:

(i) Catalytic steam methane reforming (SMR). The steam reforming is the predominant commercial technology used currently to produce syngas (H_2/CO ratio of ~3) from methane (Eq. 1). The process is highly endothermic and it is conducted in the presence of a catalyst; Ni, Fe, Co and Pt are mostly used in this process. Typical operation conditions with Ni catalysts are 1100-1300 K and 1.5-2.0 MPa. An important disadvantage is the formation of large amounts of coke.

$$CH_4 + H_2O \rightarrow 3H_2 + CO \ (\Delta H^0 = +206 \text{ kJ mol}^{-1}) \quad (1)$$

(ii) Partial oxidation of methane (POM). This process transforms either catalytically or non-catalytically CH_4/O_2 mixtures into syngas (Eq. 2). This reaction is exothermic and several orders of magnitude faster than the reforming process with water steam. The process operates at *ca.* 1000 K and high conversions (~90 %). The H_2-to-CO ratio is lower than the obtained from steam reforming. The main drawback of the process is the necessity of a costly O_2 production plant.

$$CH_4 + \frac{1}{2} O_2 \rightarrow 2H_2 + CO \ (\Delta H^0 = -36 \text{ kJ mol}^{-1}) \quad (2)$$

(iii) Autothermal reforming (ATR). This process combines both steam methane reforming and partial oxidation processes in a single reaction (Eq. 3). Thus, the heat produced during the oxidation reaction is used to conduct the reforming with water vapour. The H_2/CO ratio can be tuned by modifying the CH_4/O_2 ratio.

$$CH_4 + O_2 + H_2O \rightarrow H_2 + CO \ (\Delta H^0 = -36 \text{ kJ mol}^{-1}) \quad (3)$$

1.2.2. Syngas Cleaning

The synthesis gas produced from methane, coal and/or biomass contains typically impurities such as tars, aromatic compounds (benzene, toluene and xylenes, known as BTX), inorganic gases (NH_3, HCN, H_2S, COS and HCl), and volatile metals soot and dust.

The typical catalysts used in FTS are intrinsically very sensitive to small amounts of impurities. In commercial operation, catalysts are replaced or regenerated after a certain operational period. The definition of the gas cleaning is, therefore, based on economic considerations: investment in gas cleaning *versus* decrease of reaction rate because of the poisoning of the catalyst. Therefore, there are no specific data on maximum levels accepted for impurities in FTS feed, and the acceptable levels may be different for each FTS plant.

A concentration for the sum of the nitrogen- and sulphur-containing compounds higher than 1 ppm (volume) in the synthesis gas cannot be used. There exist two primary methods for removing ammonia from syngas: catalytic destruction and wet scrubbing (preferred in FTS plants). Ammonia can be removed together with the halides (HCl, HBr, HF, <10 ppb required) with a water washer, while H_2S is removed either by absorption or after conversion to elementary sulphur (the Claus process), although the absorption process is preferred when relatively small amounts of H_2S are present. Similar is valid for the presence of COS and HCN. These impurities are hardly removed in the gas cleaning and are captured in ZnO guard beds [14].

With respect to the organic constituents (tars and BTX) that are present in biomass-obtained syngas, there are no limits regarding the poisoning of the catalyst. However, since the syngas needs to be compressed to 2-6 MPa for FTS, the concentration of the organic compounds must be below the dew point at FT pressure to prevent condensation and fouling in the system. Since these organic compounds are potential sources of additional CO and H_2, they can be cracked to form smaller hydrocarbon chains. Tar removal is performed through cracking processes that can be catalytic (with Ni catalysts) or thermal at temperatures above 1400 K with steam or oxygen. They can be also removed at lower temperatures by scrubbing with an oil medium by means of the recently proposed OLGA technology [15].

Solid particulates must be removed almost completely (< 1 mg m^{-3}) because they foul the system and may obstruct fixed-bed reactors. The most important technologies to remove particulates include cyclone filters, barrier filters, electrostatic filter and wet scrubbers.

Table 1.1. Contaminant specifications for FTS catalysts and cleaning technologies

Contaminant	Limit	Removing technology
Soot (dust, char, ash)	0 ppb	Cyclones, metal filters.
Alkaline (halide metals)	< 10 ppb	Active coal bed.
Tar	< dew point	Thermal tar cracker; oil scrubber.
BTX	< dew point	
Halide compounds (HCl, HBr, HF)	< 10 ppb	Aqueous scrubber; active coal.
Nitrogen compounds	< 1 ppm	
NH_3		Aqueous scrubber.
HCN		Active coal bed and hydrolysis.
Sulphur compounds	< 1 ppm	
H_2S		ZnO guard bed; Claus unit.
COS		Active coal bed and hydrolysis.

With respect to the other possible constituents (CO_2, N_2, CH_4, and larger hydrocarbons, depending on the technology used in syngas production) of the FT feed, there are no strict limitations. Required specifications are determined by economic considerations. Nevertheless, an upper limit around 15 vol.% is considered as acceptable for the concentration of these inert gases in FTS, although lower concentrations are preferred if possible because the presence of inert compounds requires larger reactors and higher total gas pressures. CO_2 can be easily removed, but N_2 and the low molecular weight hydrocarbons cannot be eliminated at reasonable costs. Therefore, the presence of significant concentrations of the latter compounds should be avoided. The most common contaminants, the maximum concentration and the technology available to remove them from synthesis gas are summarized in Table 1.1.

1.2.3. Syngas Conditioning

The H_2/CO ratio can be adjusted to the optimum values required in FTS by performing a previous conditioning step consisting in the water-gas shift reaction (Eq. 4). This is a catalytic process that can be conducted either at high temperatures (650-700 K) with Fe or Cr based catalysts or at low temperatures (450-500 K) with Cu-ZnO-Al_2O_3 catalysts. Water and carbon dioxide do not poison the FTS catalysts but decrease the partial pressure of CO and H_2. Therefore, they should be removed out of the gaseous feed before entering the FTS reactor.

$$CO + H_2O \rightarrow CO_2 + H_2 \; (\Delta H^0 = -41 \text{ kJ mol}^{-1}) \quad (4)$$

1.3. Active Catalysts for Fischer-Tropsch Synthesis

A variety of metals can be used as catalysts for the Fischer-Tropsch synthesis, most importantly iron, cobalt, and ruthenium [16-19]. Ru is well-known to be the most active metal in FTS, but the high cost impedes the use of this element in large-scale applications. Therefore, Co- and Fe-based materials remain as the only feasible catalysts used in commercial scale. The final selection of the active FTS metal depends on many parameters, including the price, the source of synthesis gas and the specific fraction of hydrocarbons aimed to be prepared.

1.3.1. Fischer-Tropsch Synthesis with Co-Based Catalysts

Cobalt-based catalysts form predominantly long-chain linear alkanes from H_2/CO mixtures [20]. Catalytic active sites are constituted by metallic cobalt atoms. The specific activity, also known as turnover frequency (TOF), is not significantly influenced by Co particle size when clusters are larger than ~10 nm. However, some authors have found that the intrinsic FTS rate progressively decreases for smaller assembles of atoms [21,22]. Hence, a minimum Co size is required to form the active catalytic sites for FTS reaction, although a definitive and unequivocal explanation for this requirement has not been proposed yet.

Cobalt is typically used as relative small metal clusters (10-20 nm) deposited on an oxide support (Al_2O_3, SiO_2, TiO_2, etc.) by impregnation with Co^{2+} aqueous solutions. The supporting material provides mechanical strength and thermal stability to avoid the agglomeration of Co clusters during catalyst pretreatment steps. Moreover, the chemical nature of the support determines the reducibility of cobalt oxide to form metallic Co particles, which is reflected in the metallic dispersion. Strong Co-support interactions (as in Al_2O_3 and TiO_2) favor the formation of small cobalt oxide clusters, but at the same time, decrease their reducibility. Consequently, the high temperatures required to reduce the cobalt oxide result in catalysts with a limited number of accessible Co metal sites. On the contrary, weaker Co-support interactions (as in SiO_2) form cobalt oxides with high reducibility. These cobalt particles tend, however, to agglomerate during the thermal activation treatments, resulting thus in catalysts with a low number of exposed Co atoms. Therefore, a compromise between size and reducibility is necessary in order to optimize catalyst properties. Nevertheless, the use of presynthesized nanoparticles instead of precursor salts results in improved catalytic activity due to the lower interaction of the active phase with the support [23,24]. Moreover, recent developments in microemulsion technology have accomplished the preparation of small supported Co clusters while avoiding the dispersion-reducibility dependence inherent to very small Co clusters deposited on oxidic carriers [25].

Promoters are typically added to Co catalysts to improve the activity, selectivity and/or stability. Noble metals promoters (Ru, Pt, Rh, etc.) are often used to enhance the reducibility of the cobalt oxides particles, leading thus to catalysts with smaller Co clusters [17,26]. Metal oxide promoters (MnO_x, ZrO_2, MgO, CeO_2, etc) have been also extensively studied to increase the FTS selectivity to desired products [27,28]. These oxides change the local electronic structure of metallic cobalt, resulting in the modification of its chemisorption properties. The surface coverage of reactants is thus altered, and as a consequence, the catalysis performed by the metal is modified.

Water, the main product of the FTS reaction, is known to affect the observed rate and selectivity values with Co-based catalysts [29,30]. At low water partial pressures (low CO conversion), a positive effect of water on FTS rate and C_{5+} hydrocarbons selectivity has been found. However, high water partial pressures created by high CO conversion or steam cofeeding leads to a negative effect, probably due to the formation of a cobalt-silica mixed oxide [31]. To avoid the harmful influence of water at high conversion levels, hydrophobic cobalt-based catalysts have been prepared using silylated SiO_2 materials as a support [32].

The knowledge of the chemical state of the cobalt clusters and the species adsorbed on the surface during FTS catalysis is still challenging, preventing a complete understanding of the process. This is due to the lack of suitable characterization techniques able to determine the chemical and electronic structure of the catalyst during reaction. The use of X-ray absorption spectroscopy (XAS), a synchrotron-based technique for studying the electronic structure of matter, is promising. XAS is a powerful method of local order characterization in catalytic systems, and it has been extensively used for identification of neighboring atoms, measuring interatomic distances, and evaluation of coordination number, as well as structural and thermal disorder. XAS measurements of cobalt catalysts have been typically conducted at the K-edge of cobalt (7709 eV). This relatively high photon energy allows both *in situ* and *operando* studies about the state of cobalt atoms during catalyst preparation and chemical reactions. Bazin *et al.* [33] demonstrated the feasibility of extracting information about the structure of cobalt catalysts using L_{II} and L_{III} absorption edges (778 and 790 eV). At these photon energies (soft X-rays), the experiments need to be performed at very low pressures because of the strong absorption of the X-ray beam by air. Soft X-ray absorption edges present the important advantage of being sharp because of the long lifetime of the core states. Therefore, soft X-rays XAS is sensitive to details of the electronic structure, such as the valence level structure, symmetry, spin state, etc. In the first experiments performed by Bazin, spectra of a powder Co/SiO_2 sample were obtained in the presence of H_2, although at partial pressure values (10^{-4} Pa) significantly lower than those used in the industrial systems. Using Ambient Pressure Photoelectron Spectroscopy [34], Morales *et al.* [35] studied the influence of promoters and the support on cobalt oxide reducibility using *in-situ* XPS and soft X-ray absorption. Recent developments in the design of *in situ* cells allow to acquire soft X-rays XAS spectra in a flow of reactant gases at atmospheric pressure and temperature as high as 630 K, which is similar to the temperature used in industrial conditions [36].

Recently, several authors have reported that the turnover frequency or activity per surface site decreases dramatically when the Co cluster size reaches a critical point [21,37]. This observation was not well understood because the cobalt clusters displayed a heterogeneous shape, size distribution, and different interaction with the support. Recent advanced colloidal preparation methods have resulted in the synthesis of Co nanoparticles with diameters below 10 nm, controlled shape and narrow size distribution [38]. This allows the rigorous study of size and shape influence on reaction rate and selectivity; these effects have been traditionally masked by the wide cobalt size distribution typically obtained using more conventional impregnation methods. The low activity found with small cobalt clusters (< 10 nm) has been attributed to several causes: (*i*) cobalt oxidation by H_2O produced in the reaction; (*ii*) deposit of carbonaceous species; and (*iii*) geometrical effects. A recent study using *in situ* X-ray absorption spectroscopy as a characterization tool and cobalt nanoparticles prepared by colloidal chemistry with monodispersed size distribution concluded that the different reaction rates found with a range of Co clusters is related to the distinct ability to dissociate molecular H_2 [36]. Model catalysts prepared using cobalt nanoparticles synthesized by colloidal chemistry were used to test the influence of particle size in the catalytic performance. These cobalt nanoparticles were supported over a gold foil using the Langmuir-Blodgett technique in order to obtain two dimensional (2D) model catalysts. In addition, the Co nanoparticles were deposited on SiO_2 to thus prepare conventional Co-based FTS materials (3D). The influence of the Co cluster size on the catalytic performance measured with the 3D catalysts is similar to that published previously in the literature (decreasing turnover frequency values when metal clusters are smaller than 10 nm). Interestingly, the activation energy values calculated from Arrhenius plots are independent of Co size, suggesting the rate determining step is the same for all particle sizes. Bidimensional (2D) model catalysts were characterized *in situ* by soft X-ray absorption spectroscopy (atmospheric pressure and up to 623 K). Changes in the electronic structure of the metallic cobalt nanoparticles were followed both at room temperature and high temperature upon switching the H_2 stream and passing the H_2/CO mixture. Irrespective of the cobalt size, the Co L-edge spectra did not change after flowing CO at room temperature or heating in CO or H_2/CO mixtures. Simultaneously, near-edge X-ray absorption fine structure (NEXAFS) spectroscopy was performed to follow changes in the oxygen K-edge. The reduced cobalt nanoparticles showed no features in the O K-edge spectra. After exposing to CO at room temperature, a large absorption peak at 534.2 eV appeared, which was ascribed to molecular CO adsorbed on the cobalt surface [39]. When the model catalyst was heated in flowing H_2 at 240°C, this peak associated to CO adsorption disappeared due to the reaction with H_2 to form CH_4. If the same treatment was performed in flowing He, the CO-related peak in the O K-edge at 534.2 eV decreased because of desorption. These results showed that the active phase of Co catalysts in Fischer-Tropsch synthesis is the metallic state and that it remains unchanged during the reaction. Van Steen and coworkers suggested that smaller Co nanoparticles are more susceptible to water oxidation using theoretical calculations, resulting in lower turnover frequencies in the CO hydrogenation reaction [40]. However, in 4-5 nm SiO_2-deposited Co nanoparticles, oxidation was not observed until very high temperatures (more than 400°C), even with a H_2/H_2O ratio of 1 [41]. This is consistent with the soft XAS results, which revealed the absence of any peak assigned to cobalt oxide in the O K-edge region. Moreover, the Co L-edge peak structure was always that of Co in the metallic state during the heating in syngas.

The differences in catalytic activity as a function of particle diameter cannot be attributed to oxidation of the catalysts during reaction, as shown by the *in situ* XAS results. Experiments performed with the nanoparticles mixed with silica (3D catalysts) demonstrated that the catalytic activity was stable during the time-on-stream (TOS) studied and deactivation by carbon deposition could also be ruled out, because after the catalytic reaction tests, temperature-programmed treatment in hydrogen of all the samples did not form carbon-containing products. Sintering of the cobalt nanoparticles was also discarded after characterizing the used catalysts by scanning and transmission electron microscopy. It has been previously proposed that the cause of the different reactivity was the altered H_2 chemisorption on cobalt. Hydrogen dissociation step might require a particular geometry of a minimum of two or more adjacent empty Co sites. Indeed, it has been shown that H_2

dissociation requires at least three adjacent sites on Pd, Pt and Ru [42]. Therefore, hydrogenation of CO would require three or more adjacent sites to dissociate hydrogen and at least one more for CO adsorption, as in the H-assisted CO dissociation mechanism suggested by King *et al.* [43]. The probability of finding these larger ensemble sites should decrease rapidly for the smaller particles, thus explaining qualitatively the decrease of the turnover frequency with smaller Co particles. This model was tested by performing H_2-D_2 exchange experiments on selected samples of reduced cobalt nanoparticles supported on SiO_2 powder. As expected, the specific activity for HD production increases with temperature, but more significantly, the HD turnover frequency decreases with particle size. The results demonstrate that H_2 dissociation on Co is structure sensitive for particles with size below 10 nm.

Cobalt is not active towards the water-gas shift reaction ($CO + H_2O \rightarrow CO_2 + H_2$). Therefore, Co catalysts are preferred when natural gas-derived synthesis gas ($H_2/CO = 2$) is used. For H_2-poor synthesis gas ($H_2/CO < 2$), such as those derived from coal and biomass, Fe-based catalysts are more suitable because they are active in WGS and the ratio H_2/CO can be properly adjusted.

1.3.2. Fischer-Tropsch Synthesis with Fe-Based Catalysts

Fe catalysts are also widely used in FTS because of the low price and flexible product distribution, which includes paraffins, linear and branched olefins, and oxygenated compounds. The reaction selectivity depends on the catalyst composition (promoters) and process conditions employed.

In contrast to cobalt, which remains in the metallic state during FTS, iron-based catalysts tend to form several phases during the course of the reaction. Thus, metallic iron (α-Fe), iron oxides (hematite α-Fe_2O_3 and magnetite Fe_3O_4) and different forms of iron carbides (O-carbides: ε-Fe_2C, ε'-$Fe_{2.2}C$ and FeC_x, with carbon atoms in octahedral interstices; and TP-carbides: χ-$Fe_{2.5}C$ and θ-Fe_3C, with carbon atoms in trigonal prismatic interstices) may co-exist during reaction. Most of them have been proposed at some time as the active Fe structure for FTS [18].

The formation and composition of the different Fe phases (which determines the catalyst activity. selectivity and/or stability) depends on the process conditions, activation procedure and catalyst composition. Moreover, the relative concentration of these phases evolves during the reaction. The most active Fe phase in FTS has not been unequivocally identified yet because of the absence of effective *in situ* characterization techniques. Most recent studies propose that the formation of iron carbides is a requirement to obtain high FTS rates, although indirect measurements have been typically used to determine the identity and structure of the Fe active sites.

Fe catalysts are mostly prepared by precipitation because this metal is considerably less expensive than Co. The experimental method consists typically in the precipitation of iron hydroxides and oxides from an aqueous solution of Fe precursors. The precipitate is subsequently washed, dried and treated in air at high temperature. Finally, the catalyst is activated in H_2 or synthesis gas before performing the catalytic reaction.

Precipitated Fe catalysts tend to possess low structural strength. Consequently, the significant disintegration of the Fe particles may cause plugging, fouling, difficulty in separating the catalyst from the wax product, and loss of the catalyst. Improvements in catalyst attrition resistance can be obtained by using supported iron catalysts, although these materials display generally lower reaction rates.

Oxides (ZnO, MnO), metals (Cu, Ru) and alkali (K, Na, Cs, Rb) oxides or carbonates are typically added to Fe oxide precursors as promoters in order to improve their structural integrity or catalytic properties. Potassium is probably the most studied promoter for iron catalysts because it enhances CO chemisorption and inhibits H_2 chemisorption, affecting thus the relative surface coverage of reactants [44]. This leads to lower FTS rates, higher product molecular weight, and increased olefin content. Fe promotion with Cu has been also extensively studied. Cu decreases the temperature required during the catalyst reduction pretreatment, which may result in higher FTS rates as a result of the formation of smaller Fe crystallites [44,45]. A structural promoter may be also added to an iron catalyst to provide a large surface area for the formation and stabilization of the small metal crystallites. Thus, silica is frequently added as a binder to precipitated iron catalysts, especially for those used in fixed-bed reactors.

As described in the previous section, the synthesis gas is commonly purified to remove several substances that may act as catalyst poisons [46]. FTS catalysts are notoriously sensitive to the presence of small concentrations of sulfur-containing compounds in the synthesis gas feed, although Fe-based catalysts have been shown to be more sulfur-resistance than their Co-based counterparts. Sulfur compounds typically lead to permanent deactivation of the FTS catalyst, while N-containing compounds, as ammonia and HCN cause a kinetic inhibition of the reaction rate. In this case, Fe catalysts are also more resistant than Co-based materials.

As a summary, Table 1.2 compares the performance and properties of Fe and Co catalysts in the FTS.

Table 1.2. Comparative overview of the features and FTS performance of Fe and Co

Property	Iron	Cobalt
Activity in FTS	Lower reaction rates	Higher reaction rates
Selectivity	Small chain hydrocarbons High olefins content Oxygenated compounds	Long chain hydrocarbons Low olefins content No oxygenated compounds
Preferred syngas source	Coal or biomass	Natural gas
Water-gas shift activity	Active	Inactive
Poisons	Deactivated by H_2S	Deactivated by H_2S
Reaction temperatures	High temperatures (570-630 K)	Low temperatures (470- 530 K)
Cost	Inexpensive	Relatively expensive

1.3.3. Fischer-Tropsch Synthesis with Ru-Based Catalysts

Ruthenium is the most active metal in the CO hydrogenation reaction among Group VIII metals, producing high molecular weight paraffinic waxes at high pressure and low temperatures [47]. The typical ranges of chain growth probability (α) on Ru are 0.85-0.95, whereas on Co and Fe catalysts are 0.70-0.80 and 0.50-0.70, respectively [47]. Ruthenium catalysts form very large hydrocarbons at atmospheric pressure, which justifies the use of Ru as a catalyst in the FTS reaction, even though its high price. Furthermore, Ru-based catalysts can be used at CO/H_2 conversions higher than that obtained with Fe or Co in a typical FT reactor, leading to lower FTS fuels production costs.

Product distribution obtained with Ru catalysts is strongly affected by both the reaction conditions and catalyst composition. Thus, formation of unsaturated hydrocarbons increases with CO conversion, while the length of the hydrocarbon chain depends on the temperature, pressure, and nature of the support. As with iron and cobalt catalysts, the length of the hydrocarbon chain increases with pressure and decreases as the temperature is increased.

The specific activity of ruthenium in FTS is determined by the particle size, although unequivocal results have been found. Ruthenium catalysts with 2-7 nm particle size have been prepared using reverse micelle solutions and conventional impregnation on Al_2O_3. Particles below 3-4 nm agglomerate during FTS (481-498 K and $H_2/CO=0.8-2.0$), whereas particles of 5-7 nm clusters remain unchanged and are more active than the smaller particles [48]. More recent studies confirm that the optimum Ru size is around 7-8 nm Ru [49,50]. Nevertheless, the metal particle size effect when Ru>8 nm is not clear.

Ru catalysts can operate at high partial pressures of water. In fact, the addition of water to the feed stream has a positive effect in the activity and heavy hydrocarbon selectivity. It has been proposed that water provides a hydrogen source to form chain growth monomers and alkyl surface species [50]. Ru catalysts can also operate under other oxygenate-containing atmospheres, which is an important property to convert syngas obtained from biomass. However, ruthenium is not active in reverse water-gas shift reactions ($H_2O + CO \leftrightarrow H_2 + CO_2$), and syngas with $H_2/CO=2$ is preferred.

Due to the high cost, ruthenium catalysts are commonly prepared as small clusters supported (0.5-5.0 wt.% metal loading) on a wide range of different oxides (SiO_2, Al_2O_3, ZrO_2, TiO_2, ThO_2, etc) by impregnation. The support plays an important role in the observed activity due to electronic metal-support interactions and providing stability against metal sintering. The Ru-support interactions are associated to the formation of chemical bonds between Ru and the metal of the support, involving intermetallic compounds and/or as a result of the overlapping of the occupied *d* orbitals of Ru cations with the vacant *d* orbitals of the metal. This effect is clearly observed in the catalysts prepared over titania, displaying better catalytic performance compared to other support, an effect known as strong metal-support interaction (SMSI) [51,52].

Ruthenium catalysts show significant deactivation during FTS because the deposition on carbonaceous species on the surface. Neither oxidation of the active phase nor the agglomeration of Ru clusters under H_2/CO atmosphere seems to play important roles in the deactivation mechanism [53].

The catalytic performance of ruthenium catalysts can be modified and/or improved by the addition of different elements, such as Mn, which increases the activity and stability [54], or K, that increases the selectivity towards higher hydrocarbons. The addition of Zr and Li to Ru/SiO_2 catalysts leads to a more stable catalyst as a result of the cooperative effect between Ru, SiO_2 and these promoters [55].

Typically, Ru is added as a promoter of Fe or Co FTS. Ruthenium is added to cobalt catalysts to enhance the reducibility of CoO_x particles and increase the activity and the selectivity to heavier hydrocarbons of these catalysts [56]. Ru promoters also improve Fe oxides reducibility in bimetallic FeRu catalysts [57].

1.4. Mechanisms, Product Distribution, and Thermodynamics

1.4.1. Mechanisms of Fischer-Tropsch Synthesis

The Fischer-Tropsch Synthesis is a complex network of primary and secondary reactions. Although the FTS process has been extensively studied for more than 100 years, an unequivocal reaction mechanism supported by definitive experimental evidence is not available yet in the literature.

It is widely accepted that FTS is a polymerization reaction in which surface CH_x* species are formed *in situ* from carbon monoxide and hydrogen. It is uncertain, however, if the monomer formation proceeds *via* hydrogenation of dissociated or undissociated CO molecules. Irrespective of the formation route, CH_x* species act as monomers and chain initiators in chain growth processes to form mainly paraffins and olefins with a broad range of chain size. Chain growth occurs by stepwise addition of C_1 monomers to a surface alkyl chain. The process terminates by the addition or elimination of hydrogen to form predominantly *n*-paraffins or *1*-olefins, respectively (Figure 1.2). Chain termination can also occur by CO insertion into surface alkyl chains, leading thus to the formation of primary alcohols [16]. The overall stoichiometry of the synthesis of paraffins, olefins and alcohols is represented by the following equations:

$$\text{Paraffins: } n\,CO + (2n+1)\,H_2 \rightarrow C_nH_{2n+2} + n\,H_2O \quad (5)$$

$$\text{Olefins: } n\,CO + 2n\,H_2 \rightarrow C_nH_{2n} + n\,H_2O \quad (6)$$

$$\text{Alcohols: } n\,CO + 2n\,H_2 \rightarrow C_nH_{2n+2}O + (n-1)\,H_2O \quad (7)$$

The *1*-olefins can be readsorbed onto chain growth sites and initiate new surface chains that continue to grow and form eventually larger hydrocarbons.

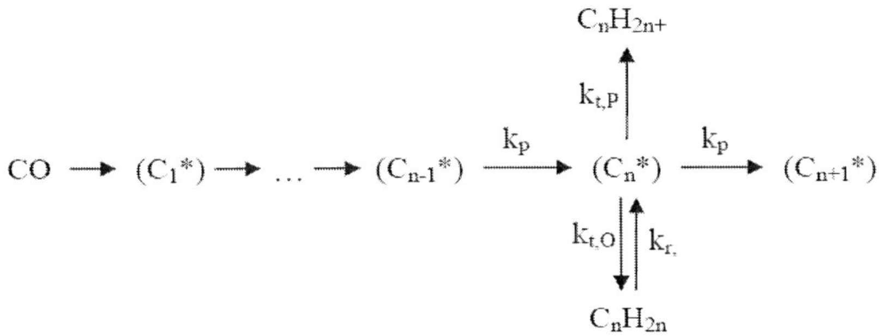

Figure 1.2. Chain growth and termination reactions in Fischer-Tropsch Synthesis (k_p, $k_{t,P}$, $k_{t,O}$ and k_r are the rate constants of chain propagation, termination as paraffins and olefins, and readsorption).

Several reviews concerning the FTS mechanism have been published in the literature [18,19,47,54,58] The polymerization FTS occurs through the following steps: (*i*) reactant adsorption; (*ii*) chain initiation; (*iii*) chain growth; (*iv*) chain termination; (*v*) product desorption; and (*vi*) readsorption and further reaction. Most of the FTS mechanisms proposed in the literature may be divided into three main classes:

1. Carbide mechanism. This mechanism, based on the early proposal by Fischer and Tropsch [5], remains currently as the most accepted route for hydrocarbon formation from CO/H_2 mixtures on Fe, Co and Ru catalysts. This pathway involves CO adsorption and subsequent dissociation, followed by stepwise reaction between carbon (C*) and hydrogen (H*) adatoms to form the reaction monomer CH_2* species. These species polymerize afterwards by successive addition of C_1 units to yield larger surface alkyl chains. The presence of methylene species has been identified by using isotopic-tracer techniques on Ru/SiO_2, Ni/SiO_2, Ru/Al_2O_3 and Fe/Al_2O_3 catalysts [59-62] . Chain growth terminates by either *β*-hydrogen abstraction to form olefins or hydrogen addition to yield paraffins [20]. Several authors have considered also a "*modified*" FTS carbide mechanism where CH_2* species are formed by hydrogen assisted CO dissociation pathways [43,63-65]. It has been demonstrated that the addition of H* to adsorbed CO* species to form CH_xO* (x = 1,2) species is energetically preferred over the unassisted CO* dissociation. These species dissociate subsequently to form surface CH* groups, which are then hydrogenated to the FTS monomers (CH_2*). The reaction selectivity is determined by the relative rates of chain growth and termination pathways, while the rate of formation of the monomer building blocks dictates the overall CO consumption kinetics. This reaction pathway does not explain, however, the formation of oxygenated hydrocarbons (mainly alcohols), which can be formed as FTS by-products in relatively significant amounts, depending on the reaction conditions and catalyst used.

2. Hydroxy-carbene mechanism. This proposal suggests the formation of hydroxy-carbene (CHOH*) species as the reaction intermediates and starting point of chain propagation steps. The CHOH* species are formed by the partial hydrogenation of undissociated adsorbed carbon monoxide, similarly to the "*modified*" carbide mechanism described above. However, the carbon-carbon bound formation occurs here via the condensation of two hydroxy-carbenes by elimination of water. This model explains successfully the formation of both oxygenates and hydrocarbons [66]. This mechanism is supported by the participation of co-fed alcohols to CO/H_2 mixtures in chain growth processes. It has been noted, however, that the adsorption of co-fed alcohols and the involvement of the resulting intermediated in chain growth steps do not necessary require O-containing species [67]. Moreover, the formation of C-C bonds from two electrophilic hydroxy-carbene species is not totally indisputable.

3. CO-insertion mechanism. The main reaction intermediate in the CO insertion mechanism, proposed by Pichler and Schulz [68], is a metal carbonyl species, which is partially reduced by adsorbed hydrogen to metal-alkyl intermediates. These can undergo afterwards a variety of reactions to form acids, aldehydes, alcohols and hydrocarbons. The chain growth process to form larger hydrocarbons can be explained by the insertion of an associatively adsorbed carbon monoxide into the metal-alkyl bond. The termination of the chain growth process occurs by desorption of products from the catalysts surface.

One important common feature shared by the previous three mechanisms is the presence of a single key intermediate. However, none of these pathways is capable of predicting the whole product spectrum observed with the typical metals used in FTS. This has led researchers to assume that the Fischer-Tropsch mechanism likely involves two or more key intermediates [58]. In this respect, it has been proposed that a CO-insertion mechanism is responsible for the formation of oxygenates, while hydrocarbons are formed via the O-free carbene mechanism (Figure 1.3). It should be noted that the surface monomers (CH_x) can be formed via unassisted and H-assisted CO dissociation pathways. It is emphasized that a large variety of O-, H-, and C-containing species are present on the catalyst surface that all may be involved in the FTS mechanism. Recently, Ojeda *et al.* have provided experimental and theoretical evidence for the significant role of H-assisted pathways in kinetically-relevant CO dissociation steps on both Fe and Co catalysts [63]. Such pathways lead to the preferential oxygen rejection as water via the direct formation of OH* precursors through dissociation of H_xCO species; they represent the exclusive CO activation pathways on Co surfaces. H-assisted pathways occur concurrently with unassisted CO dissociation on Fe-based catalysts; the latter lead to the exclusive rejection of O-atoms from CO as CO_2.

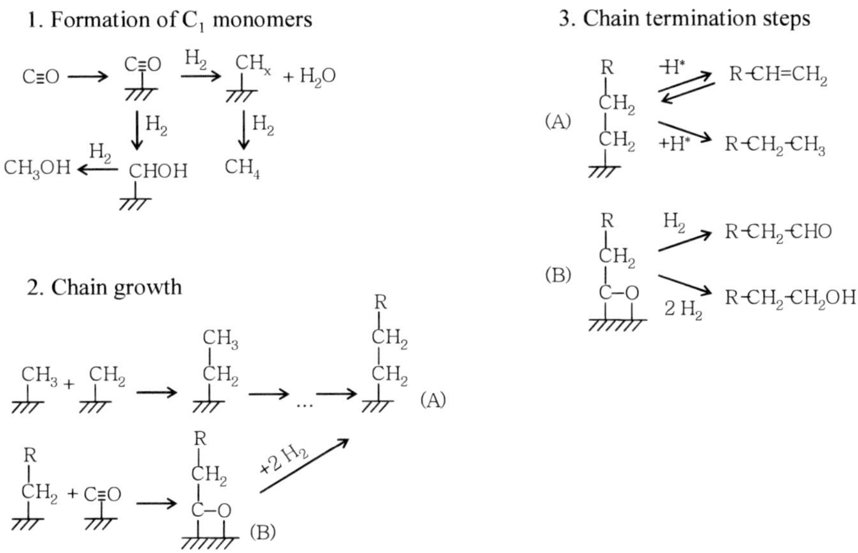

Figure 1.3. Reaction mechanism for the Fischer-Tropsch Synthesis (adapted from [42]).

1.4.2. Products Distribution

FTS product distribution is typically described by equations developed originally for polymerization kinetics. According to this model, also known as Anderson-Schulz-Flory (ASF), the product distribution in FTS is determined by the following expression:

$$W_n = n \cdot (1-\alpha)^2 \cdot \alpha^{n-1} \quad (8)$$

where W_n is the fraction of the carbon atoms within chain containing n carbon atoms. Figure 1.4 shows the selectivity to different FTS products as a function of the chain growth probability.

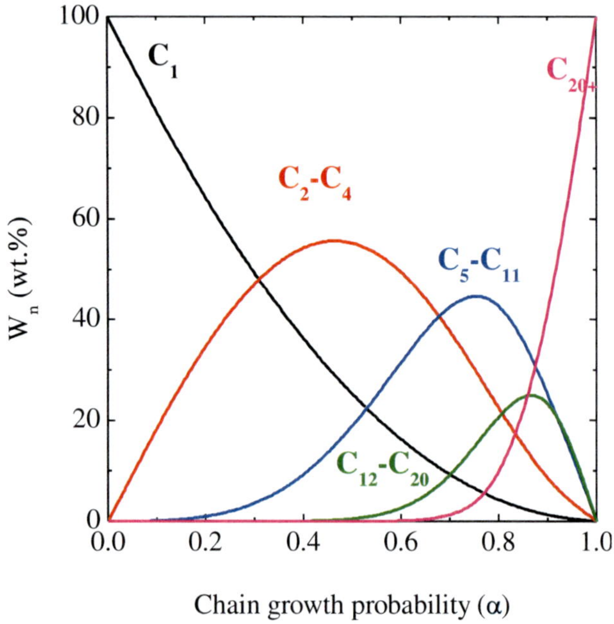

Figure 1.4. Selectivity to different FTS reaction products as a function of the chain growth probability (α).

In the ASF kinetics, the chain growth probability (α) is independent of chain size. However, FTS products seldom follow ASF kinetics throughout the entire molecular weight range. In most cases, α depends on chain size, increasing for the larger hydrocarbons. Thus:

$$\alpha_n = \sum_{i=n+1}^{\infty}\phi_i \Big/ \sum_{i=n}^{\infty}\phi_i = \frac{r_{p,n}}{r_{p,n}+r_{t,n}} \tag{9}$$

where ϕ_i denotes the molar fraction of the product with n carbon atoms, while $r_{p,n}$ and $r_{t,n}$ are the rates of chain growth and termination, respectively. It has been previously proposed that these deviations reflect the increasing transport-limited removal of reactive olefins from catalyst pellets as the chain size increases.

The range of α is dependent on the reaction conditions and catalyst type. Dry reported typical ranges of α on Ru, Co, and Fe of 0.85-0.95, 0.70-0.80, and 0.50-0.70, respectively [69].

An alternative approach proposed by Iglesia *et al.* [16,70] consists in describing FTS products in terms of individual chain termination probabilities for each chain size (β_n):

$$\beta_n = \phi_n \Big/ \sum_{i=n+1}^{\infty} \phi_i = \frac{r_{t,n}}{r_{p,n}} = \frac{1-\alpha_n}{\alpha_n} \tag{10}$$

This approach allows chain growth kinetics to vary with chain length. The total termination probability ($\beta_{T,n}$) is a linear combination of the values for the individual termination steps (olefins, paraffins and readsorption):

$$\beta_{T,n} = \beta_{O,n} + \beta_{P,n} - \beta_{r,n} \tag{11}$$

where the individual termination probabilities for olefin formation ($\beta_{O,n}$), paraffins ($\beta_{P,n}$) and readsorption ($\beta_{r,n}$) are obtained upon substituting the mole fraction of each hydrocarbon species in the numerator of Eq. 10.

The product distribution is also affected by secondary reactions occurring during FTS catalysis. All products formed by desorption from a chain growth site are primary products. Secondary reactions change FTS selectivity by chemical transformations of the primary products on a second catalytic function. The most important secondary reactions for *1*-olefins include: (*i*) isomerization to internal olefins; (*ii*) cracking and hydrogenolysis; (*iii*) hydrogenation to paraffins; (*iv*) insertion into a growing hydrocarbon chain; (*v*) chain initiation. In this respect, there has been some debate about whether paraffins are primary products formed via the hydrogen-assisted termination of growing hydrocarbon chains or whether they are exclusively formed via the adsorption of olefins on secondary hydrogenation sites. Nevertheless, it is mostly accepted that both paraffins and olefins are primary FTS products. Secondary reactions, where the carbon number of the primary product is altered (cracking, hydrogenolysis, chain insertion, and chain initiation), obviously affect the product distribution. These reactions are partially responsible for the experimentally observed deviations of the ASF distribution. FTS products seldom follow the predicted distributions throughout the entire molecular weight range [16]. Relatively high contents of methane are typically reported in the literature. The methane termination probability parameter appears to be between 5 and 20 times larger than the termination probability to other paraffins [18]. Different catalytic sites for the methanation reaction, and heat and mass transfer limitations are commonly reported in literature as possible reasons for the high yields to methane. Ethene and propene show higher reactivity and larger readsorption constants than other olefins, being reflected in unexpectedly low selectivities to C_2 and C_3 hydrocarbons. Iglesia *et al.* [16,20] have proposed that non-ASF distributions often reflect the transport-limited removal of reactive olefins from catalyst pellets, consistent with the similar effects of bed residence time and molecular size on chain growth probability and product functionality.

1.4.3. Thermodynamics

Under the typical reaction conditions used in FTS, the formation of hydrocarbon is not significantly affected by chemical equilibrium. Nevertheless, thermodynamics provides useful information about the different competitive reactions that occur simultaneously. The thermodynamic probability of the formation of the different products can be obtained assuming that the selected individual reaction occurs independently [71]. Figure 1.5 shows the change of the Gibbs free energy (ΔG^0) for the formation of different FTS products (methane, ethane, ethane, n-butane, methanol and ethanol).

Table 1.3 shows ΔG^0 values for the formation of C_1-C_3 products at 500 K. The formation of methane is thermodynamically preferred in the temperature range 300-1000 K. At typical reaction temperatures (450-600 K), the formation of alcohols (especially methanol) is not favoured from thermodynamics because the positive values for the change of the Gibbs free energy. Within the group of paraffins, the formation probability decreases with increasing carbon number; an inversely proportional trend was obtained, however, for the olefinic hydrocarbons. An increasing reaction temperature leads to a shift of the product spectrum towards the olefinic and oxygenated hydrocarbons with a simultaneous decrease of alkane formation. In summary, the probability of formation of products decreases in the following order: methane > alkanes > alkenes > oxygenates. It should be noted that this order is based on thermodynamic calculations without considering the effects of the reaction kinetics control. The kinetic limitations, which may determine a catalytic system, can have important consequences on the order of formation probability predicted from thermodynamics.

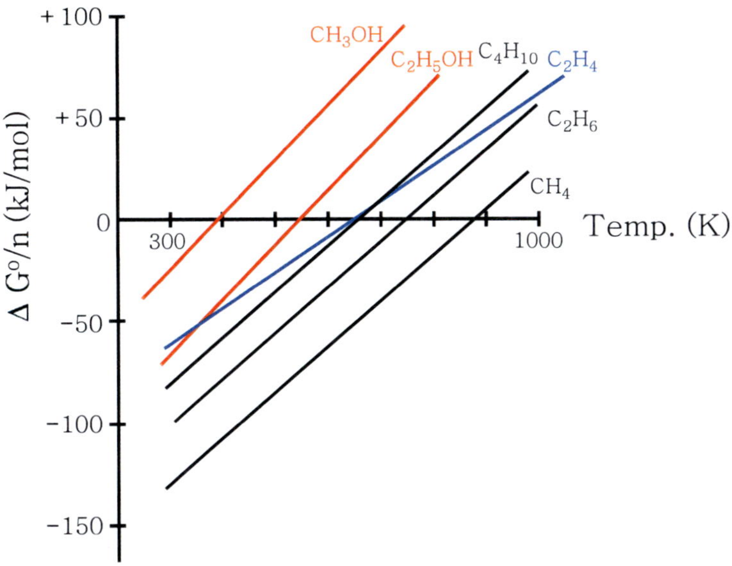

Figure 1.5. Change of the Gibbs free energy (ΔG^0) for the formation of different FTS reaction products.

Table 1.3. Change of the Gibbs free energy (ΔG^0) for the formation of C_1-C_3 FTS products at 500 K

Carbon number	Compound	$\Delta G^0/n$ (kJ mol^{-1})
1	CH_4	-92.0
	CH_3OH	+22.2
2	C_2H_6	-61.0
	C_2H_4	-23.0
	C_2H_5OH	-14.2
3	C_3H_8	-51.8
	C_3H_6	-32.2
	C_3H_7OH	-33.4

1.5. Operating Conditions and Reactors

Typical reaction conditions in FTS are temperatures comprised between 470 and 620 K and total pressures in the range 2-6 MPa. The FTS typically operates in one of the two following temperature regimes [72]: (*i*) *Low temperature Fischer-Tropsch (LTFT).* This process operates with temperatures ranging between 470 and 530 K with either Fe or Co catalysts to form predominantly high molecular weight hydrocarbons (waxes). (*ii*) *High temperature Fischer-Tropsch (HTFT).* This process uses a Fe catalyst between 570 and 630 K to form mainly gaseous hydrocarbons.

In principle, different reactor technologies are suitable for performing the highly exothermic FTS (enthalpy of reaction 170 kJ mol^{-1}). Each type of process operates with different reactors. Moving bed reactors are typically employed in the HTFT, whereas fixed bed multitubular or slurry bubble column reactors found application in LTFT. These reactors are considered as the most advanced option and are widely used in industrial application. Thus, fixed-bed multi-tubular rectors are used in the Shell Middle Distillate Synthesis (SMDS) process (tubular reactors with a diameter of 2.54 cm and a length of 12.2 m). These reactors are easy to manipulate and to design because the parallel tubes behave very similarly. This reactor presents, however, several disadvantages, such us high pressure drop, low catalyst utilization and insufficient heat removal. The pressure drop can be minimized by decreasing the catalyst pellets size (1-3 mm), although this leads to a negative effect on products selectivity. The formation of hot zones inside the reactor can be partially avoided by adding liquid products at the reactor inlet, operating thus in the trickle-flow regime. However, severe mass transfer limitations limit this approach. Another possibility for heat removal in fixed bed reactors is the application of sufficiently high gas recycles with external heat removal under adiabatic reactor operation.

The other type of reactors, slurry bubble column rectors wits suspended catalyst (developed by Sasol), manage reasonable well heat transfer issues. Moreover, these reactors use catalyst powders with dimensions of 10 to 200 μm. Thus, the influence of internal mass transfer resistances are negligible and optimal activity and selectivity can be achieved. The slurry reactor appears to be the most efficient system for production of light olefins and gasoline and diesel production. Nevertheless, the slurry bubble reactors present important problems related to the catalyst separation. Moreover, the scaling-up could be a serious drawback to this type of reactors, which otherwise, found broad acceptance for low temperature FTS.

Advantages and disadvantages of fixed bed and bubble column reactors are summarized in Table 1.4 (adapted from reference [72]).

Table 1.4. Advantages (+) and disadvantages (-) of FTS reactors

	Fixed Bed	Bubble Column
Pore diffusion	-	+
Amount of catalyst	+	-
Mass transfer	+	-
Heat removal	-	+
Catalyst exchange	-	+
Catalyst attrition	+	-
Scale up	+	-
Cost	-	+

Nowadays, different approaches are under consideration in order to improve reactor performances in FTS. Thus, the use of alternative catalysts geometry is a preferred option. Honeycomb monolith catalysts (large number of identical, parallel channels with a high cell density) present several advantages, such us high external surface area, easy scale up and low pressure drop. Furthermore, they improve very significantly mass transfer between gas, liquid and the catalyst phase because of the high surface area. However, liquid recirculation is necessary to maintain high liquid flow rates required to maintain the slug-flow or Taylor-flow regime inside the capillaries.

The application of foams for FTS appears also as a promising alternative because of satisfactory axial and radial heat and mass transfer combined with a low pressure drop. Other approaches involve the use of novel reactors; the so-called microstructured reactors show adequate features because the presence of a large number of parallel channels leads to superior heat transfer and mass transfer properties. In fact, even for high exothermic reactions such as the FTS, isothermal operation is possible with microstructured reactors.

Finally, membrane reactors also appear as an option to be investigated in the future. These reactors distribute the reactants through a membrane, which is reflected in an improved temperature control. Furthermore, the reaction selectivity can be manipulated because it depends on the H_2/CO ratio. Thus, a distribution of H_2 in a stream of CO can lead to a decrease in methane selectivity and to a higher yield to long chain hydrocarbons.

1.6. Upgrade of Fischer-Tropsch Synthesis Products

Fischer–Tropsch synthesis is a promising method to produce synthetic liquid fuels from biomass, natural gas or coal. However, the reaction products distribution follows the statistical Schulz–Flory function. This means that the selectivity to gasoline hydrocarbons (C_5–C_{11}) and middle distillates (C_{10}–C_{20}) is limited to maximum values of ca. 48 and 40 %, respectively [73]. Moreover, linear olefins and paraffins are predominantly formed [74]. These compounds exhibit low octane numbers, and consequently, extensive transformation of the FTS products is required to improve the octane number. An interesting approach to overcome the multi-stage processing required to increase the yield and quality of the direct FTS fuels is the *in situ* upgrade by using hybrid catalysts comprising a FTS catalyst and an acid or bifunctional metal/acid co-catalyst. In this way, primary FTS products (mainly *1*-olefins and long chain *n*-paraffins) are converted *in situ* into the desired gasoline components (i.e. high-octane isoparaffins) via oligomerization, cracking, and isomerization reactions occurring on the Brønsted acid sites of the co-catalyst. Sulfated zirconia has been selected in several works as a co-catalyst [74], although the catalyst deactivated rapidly because of sulfate reduction, even at very low CO conversion values. Zeolites have been widely explored as acid co-catalysts for in situ upgrading of FTS products because of their tunable acidity (number and strength of the acid sites), high hydrothermal stability, and the possibility of introducing shape selectivity effects [75]. Thus, various types and combinations of FTS catalysts and zeolites have been studied during the last decades. The FTS catalysts include iron [76], iron/manganese [77], cobalt, cobalt/manganese [78], and iron/cobalt [79], The zeolites investigated include mordenite, erionite, ZSM-11, ZSM-12, and ZSM-5 [79,80]. The identity of the hydrocarbons formed in the hybrid catalysts is determined by both the FTS active metal (Fe or Co) and the zeolite co-catalyst. Thus, Co-zeolite catalysts produces predominantly branched paraffins because Co forms *n*-paraffins that are isomerized and cracked on the zeolite acid sites. The zeolite topology was found to be determinant when Co-based catalysts are used. Indeed, the accessibility of the n-paraffins to the acid sites is determined by the particular zeolite structure, and this has a larger influence on product distribution and composition than the strength of the acid sites [81]. Thus, zeolites displaying acid sites accessible through 12-membered rings (ZSM-12) and 10-membered ring apertures (ZSM-5, ZSM-11) were more selective towards light products than ZSM-34, which comprises 8-membered ring channels through which diffusion of the straight chain hydrocarbons formed on the Co-based catalyst is hindered. Even for 10-membered ring zeolites, most of the acid-catalyzed reactions involving waxy hydrocarbons take place on the acid sites located on the external surface and/or close to the pore entrances [81,82]. Thus, the yield to high-octane gasoline components (mainly isoparaffins) increased with decreasing the HZSM-5 crystallite size. These results clearly revealed that diffusion-limited access of the *n*-paraffins to the acid sites located within zeolite channels is the key issue in the use of hybrid catalysts when Co-based materials are used as FTS catalysts.

The most serious obstacle for the use of hybrid catalysts in large scale plants is related to deactivation of the acid zeolite. In the case of Co-based catalysts, deactivation was observed to be caused by accumulation of carbonaceous species (mainly polyaromatic-type coke) on the external zeolite surface and inside the micropores. The deactivation rate was found to be larger for large-pore (12-membered ring apertures) and zeolites (USY, beta, mordenite) as compared to the medium-pore HZSM-5 because formation of polyaromatic type coke inside the 10-MR pores was hindered [83]. Combining activity and deactivation features, HZSM-5 seems to be an appropriate zeolite as acid co-catalyst for the bifunctional process when Cobalt is used as FTS catalyst.

The deactivation of the zeolite co-catalyst can be retarded by adding a noble metal (Pt, Pd) to the zeolite or introducing a third component (Pt or Pd supported catalyst) in the hybrid system. The presence of the noble metal appears to be crucial to avoid the formation of dehydrogenated coke molecules. However, unreacted CO can poison these noble metals, decreasing thus the stability of the hybrid system. In this sense, the use of Pd (more resistant to CO compared to Pt) and/or a dual layer configuration instead a physical mixture improve the catalyst stability because CO pressure is lower in the bottom layer.

Several groups have investigated the use of cobalt clusters deposited on H-ZSM-5 [84,85]. These catalysts showed a high selectivity to isoparaffins but low CO conversion rates because of the low reducibility of cobalt oxide structures. Other zeolites (USY, ZSM-5, and MCM-22) have been also used as supports for Co [86].

When iron-based catalysts are used, aromatics can be formed via oligomerization, cyclization, and hydrogen transfer reactions of *l*-olefins occurring on the zeolite acid sites. Fe catalysts operate typically at higher temperatures compared to Co, and therefore, the extent of potential secondary reactions on zeolite is more significant. On the other hand, alkali metals are often added to Fe to improve catalytic activity and selectivity. Migration of alkali from the Fe-promoted catalyst to the zeolite and subsequent poisoning of the acid sites is an important concern. As with Co-based catalysts, HZSM-5 seems to be the most appropriate option as zeolite because: (*i*) shape-selective properties limits the formation of hydrocarbons above the gasoline range; (*ii*) the highly acidic nature gives a high activity for the acid-catalyzed reactions (*iii*) resistance to coking; and (*iv*) high stability under hydrothermal conditions. The combination of an alkali-promoted iron catalyst and an acidic HZSM-5 co-catalyst has been extensively used in the literature [87,88]. The possible alkali migration processes can be avoided by physically separating the two catalysts into different beds. In a typical batch reactor, alkali transfer between separate iron catalyst and zeolite particles is also expected to occur during collisions between the particles. The extent of this migration depends on the overall time of contact between the particles and is a topic for further experimental investigation.

Deactivation of the zeolite counterpart can occur during FTS catalysis. The deactivation rate of the zeolitic co-catalyst is related to the Al/Si content. Thus, HZSM-5 with high Al/Si ratios deactivates more rapidly than HZSM-5 with low Al/Si ratio.

The secondary reactions on the zeolite acid sites can be improved by using bifunctional core/shell catalysts where core and shell components catalyze the different reactions [89,90]. Thus, a membrane layer containing H–β zeolite that directly enwrapped a Co/Al_2O_3 FTS catalyst pellet improves the collision probability between intermediates and active sites significantly. Hydrocarbons diffuse through the zeolite membrane to leave the catalyst, and those with a straight-chain structure have a chance of being cracked at the acidic sites of the zeolite. Since the hydrocarbon diffusion rate in the membrane depends on the chain length, the longer the chain length is, the longer the hydrocarbons will remain inside the zeolite, and the higher the chance that they will be cracked, which leads to a high isoparaffin/*n*-paraffin ratio and a narrow product distribution. This core/shell catalyst ($Co\text{-}Al_2O_3$/H–β zeolite) suppresses completely the formation of C_{12+} hydrocarbons and leads predominantly to the formation of middle isoparaffins products. Deactivation studies might be addressed in the future to ensure a competitive and efficient application of this type of catalysts.

Chapter 2

IRON CATALYSTS FOR FISCHER-TROPSCH SYNTHESIS

T. Herranz and F.J. Pérez-Alonso
Department of Structure and Reactivity, Institute of Catalysis and Petrochemistry (CSIC); C/ Marie Curie 2, 28049 Madrid, Spain

2.1. PREPARATION OF FE-BASED CATALYSTS

Many different methods have been extensively studied for the preparation of Fe-based Fischer-Tropsch Synthesis catalysts. The earliest catalysts, prepared by Fischer and Tropsch, consisted of fused iron treated with alkali [5]. These catalysts were prepared by melting iron ore in the presence of one or more promoters. The resulting catalyst precursor, composed predominantly of magnetite (Fe_3O_4), possesses a very low surface area (1-15 $m^2 g^{-1}$). Although this type of catalyst is inexpensive, the observed fast deactivation hinders its use in a commercial scale. An alternative approach aiming to prepare catalyst with high surface area consists in the pyrolysis of a Fe precursor, for instance, iron carbonyl compounds. These materials show surface area values as high as 300 $m^2 g^{-1}$ with a medium particle size of *ca.* 3 μm. The material produced in this manner is formed by highly dispersed particles. Nevertheless, some particles are agglomerated during initial FTS turnovers and the high surface area is decreased in some extent.

Currently, the most common method used to prepare Fe-based FTS catalysts is *via* aqueous precipitation (Figure 2.1, left) or hydrolysis of Fe^{2+}

and/or Fe^{3+} salts (nitrates, chlorides, acetates, etc.) [91,92]. Typically, the iron solution is treated with an aqueous solution of NH_4OH or $(NH_4)_2CO_3$, leading to the precipitation of the Fe ions as oxyhydroxides (FeO_xH_y). The resulting precipitated material is often washed with deionized water, although alcohols, which have lower surface tension than water, have been also used in order to minimize pore pinching during subsequent drying, forming thus solids with higher surface area values [93]. The dried iron precursor is subsequently decomposed to hematite (Fe_2O_3) by treating in air at high temperature (600-800 K).

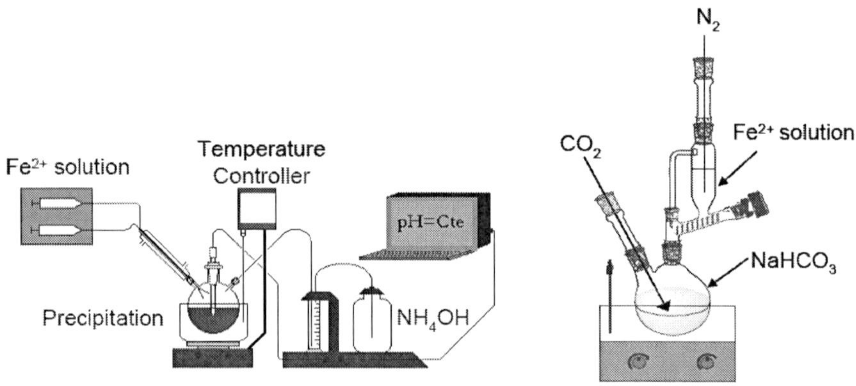

Figure 2.1. Preparation of FTS Fe catalysts by conventional (co)precipitation with a pH-state (left) or a chemical buffer (right).

The physical and chemical properties of the solids prepared by the aqueous precipitation methodology are influenced by a high number of experimental parameters, including precipitating agent, solution concentration, precipitation temperature, pH, pretreatment temperature, ageing and drying conditions, etc. When monometallic catalysts are prepared, the precipitation of iron ions is simple and fast. However, the situation becomes more complicated when bi- or multi-metallic catalytic precursors need to be synthesized. In this case, the precipitation conditions need to be controlled more carefully in order to obtain a homogeneous distribution of the different components. For instance, pH must be adjusted and controlled using a pH-state or a chemical buffer (Figure 2.1, right). Thus, the carbonate/bicarbonate buffer can be used by adding a solution of sodium bicarbonate ($NaHCO_3$) and bubbling CO_2 through the solution in order to keep constant the pH around 8.5 when iron nitrates are employed as precursors [94,95].

A controlled precipitation or coprecipitation method occurring inside aqueous micelles (microemulsion) has been recently developed to prepare Fe-based catalysts [96]. A microemulsion is an optically transparent and thermodynamically stable mixture constituted by an organic phase and an aqueous solution stabilized by a surfactant [97]. The microemulsion is called water-in-oil (w/o) when the aqueous solution is the minority phase (Figure 2.2). In contrast, the oil phase is the minor phase in the oil-in-water (o/w) microemulsions. The microscopic structure of a microemulsion w/o consists in micelles of aqueous phase surrounded and stabilized by the polar head of surfactant molecules. The use of microemulsions as nanoreactors has been generalized in order to synthesize nanoparticles of controlled size [98]. These nanoparticles can be afterwards deposited on a high area inorganic support [99]. The use of the microemulsion technology is an ideal technique for the preparation of materials containing two (or more) metallic or oxide phases because the different species are homogeneously mixed within the micelles, rendering therefore solids with high internal homogeneity and an optimal interaction between its constituents [98]. Other advantages of the materials prepared by microemulsions are the high surface area and good stability [100,101]. This methodology has been proved to yield more active and selective catalysts compared to more conventional procedures because of the resulting higher surface area of the samples and the higher extent of interaction between components in the case of bimetallic catalysts [100-103].

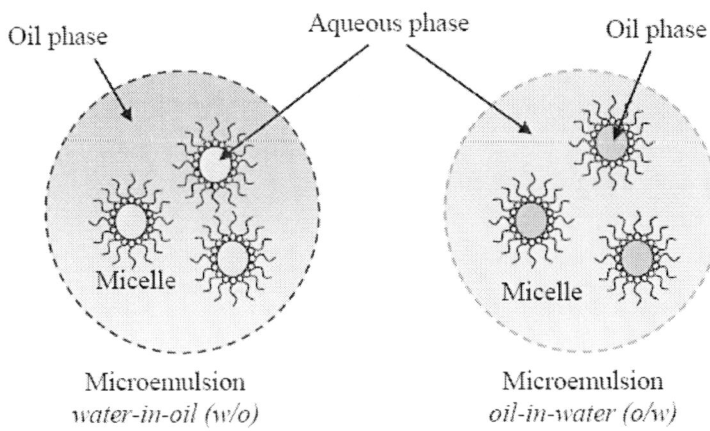

Figure 2.2. Microscopic structure of water-in-oil and oil-in-water microemulsions.

In contrast to other metals, Fe-based catalysts for FTS are usually used as unsupported materials. One of the major drawbacks related to the utilization of massive unsupported catalysts is the physical degradation, producing catalyst fines as a consequence of the volume changes that occur during the FTS reaction. These fines either will plug the fixed-bed reactors, generating a large pressure drop, or will difficult catalyst separation in slurry reactors [104,105]. The development of appropriate supports and/or binders, which increase attrition resistance of the samples, improves the catalyst life time. High surface are oxides like silica (SiO_2), alumina (Al_2O_3), titania (TiO_2), magnesia (MgO), manganese oxide (MnO) and zirconia (ZrO_2), are among the supports most frequently used. Several advantages of supported iron catalysts, such as an improved catalytic stability and lower deactivation rate, have been reported [106]. Nevertheless, catalysts containing a binder or support usually display lower activity than the unsupported counterpart. This has been attributed to the development of strong metal-support interactions that affect the reducibility of the iron phases during the pretreatment step. Thus, it has been proposed that wustite (FeO, metastable phase) is formed in the case of supported catalysts. FeO is inactive in FTS because it cannot be transformed into iron carbides, which is thought to be the active species in FTS. This results in lower catalytic activities when using supported Fe materials [107]. The wustite Fe phase has been also detected in samples containing doping agents that difficult iron reduction [108]. The use of the adequate preparation conditions is of critical importance to synthesize Fe catalysts with high dispersions and optimal reducibility [109].

More advanced and sophisticated preparation methods aiming to obtain high surface areas and well-crystallized iron oxide nanostructures have been also proposed by forming solid solutions at high temperature solution following methodologies similar to those used for preparing semiconductor quantum dots. Thus, maghemite (γ–Fe_2O_3) has been prepared by thermal decomposition of an iron complex in octylamine [110]. Oleic acid and trimethylamine oxide were used during the synthesis to control the growth of the particles and to provide a protective surfactant capping layer [111]. Similarly, magnetite (Fe_3O_4) has been synthesized using iron acetylacetonate as precursor in a mixed solution of oleic acid, 1,2-hexadecanediol, oleylamine and phenyl ether [112,113]. Other preparation techniques described in the literature are based on ultrasound and laser heating [114]. Rice *et al.* [115] utilized a laser to induce iron carbonyl present in a stream of ethene to form iron carbide catalysts. This preparation method is, however, difficult to scale to a commercial level because the production cost is not competitive at all with those materials prepared by aqueous precipitation approaches.

The solvated metal atom impregnation (SMAI) method allows the preparation of Fe-based FTS catalysts with precise control of the size of crystallites in the nanoscale range [116]. This method consists in the decomposition of an iron complex over a high area support in inert atmosphere. The reactivity of Fe nanoparticles prepared in this way is, however, similar to that shown by supported iron catalysts prepared by conventional impregnation [117]. Moreover, a rapid deactivation process occurs when contacted with the reactants stream, probably due to the sintering of the small Fe crystallites.

Model iron-based catalysts prepared by metal evaporation in ultrahigh vacuum have been also synthesized in order to perform basic research. For example, Yubero *et al.* [118] prepared thin films of hematite grown on silicon wafers by either Fe evaporation in an oxygen atmosphere or ion beam induced chemical vapor deposition (IBICVD). In the first case, metallic iron was evaporated (rate ~1 Å material min^{-1}) by joule heating through a tantalum wire where a high purity (99.5%) Fe filament was grasped. In the IBICVD method, $Fe(CO)_5$ vapor was passed over a silicon substrate which is bombarded by O^{2+} ions that decompose the iron compound [119].

2.2. Promoters for Fe Catalysts

The addition of different promoters to Fe-based catalysts is a common approach for the development of materials displaying an improved catalytic performance. Thus, the presence of different chemical elements can enhance the activity and/or selectivity of the active sites. The exact role of the promoters is usually discussed in terms of electronic and structural effects. In general, structural promoters increase the number of active sites of the catalyst, while electronic promoters increase the intrinsic reactivity (turnover frequency, TOF) of these active sites [120]. Typical elements used as promoters for Fe-based catalysts include alkaline, alkaline-earths, Cu, Mn, SiO_2, Al_2O_3, etc.

Alkaline elements (potassium is the most studied) act as chemical promoters by modifying the adsorption energetics of the reactants (H_2 and CO) onto the active sites. The observed effect of these elements on the catalytic behaviour of Fe catalysts has been explained as a consequence of the Fe ability to withdraw electronic density from potassium, resulting in a strengthening of the Fe-CO bond [121]. The corresponding higher surface coverage of the adsorbed CO species (θ_{CO*}) results in a lower relative coverage of hydrogen species (θ_{H*}). As a consequence, the probability of chain termination pathways by hydrogen addition to form paraffins decreases, being reflected in a higher selectivity towards olefins, high molecular products and the depletion of methane formation. However, it is still unclear the mechanism involved in the interaction between the alkaline elements and the Fe active phase (carbide iron species) in order to produce the atomic contact required for these electronic effects. Alkaline elements also affect the FTS reaction selectivity through secondary pathways. Branched paraffins are mainly formed via isomerization reactions of linear hydrocarbons catalyzed by the acid sites present on the support surface. Potassium (or other alkaline element) titrates these acid sites, decreasing thus the formation of non-linear FTS products [45,66,122]. Moreover, alkaline promoters increase the selectivity to CO_2 *via* water-gas shift reactions.

The amount of alkali added to the Fe catalyst determines significantly the extent and the consequences of the promoter effect. Although contradictory reports can be found in the literature, catalyst activity increases generally with increasing promoter loadings, reaches a maximum at a certain content, and then declines with further additions [45,122]. The experimental conditions also affect considerably the observed effect when promoters are added. Raje *et al.* [123] carried out a systematic study about the effect of K promoter on the FTS activity with Fe catalysts. They found that the relationship between potassium loading and FTS rate depends on the space velocity, and consequently, on synthesis gas conversion level. Thus, the FTS activity decreases with potassium loading at low conversions. At intermediate conversions, a maximum in FTS activity with potassium loading is observed, while potassium slightly enhances the reaction rate at high CO conversions levels. These effects are explained considering the influence of WGS reaction in all conversion regimes. The overall FTS activity is independent of the WGS reaction at low synthesis gas conversions. In this situation, K acts as a catalyst poison and FTS activity decreases with promoter loading. As the synthesis gas conversion increases, the partial pressure of H_2 is significantly decreased in the reactor. Under these conditions, the participation and contribution of the H_2 formed *in situ* by WGS pathways becomes more important. The extent of WGS reaction increases with potassium loading. In summary, the maximum FTS activity is found either: (*i*) with intermediate potassium loadings and synthesis gas conversion values; or (*ii*) high potassium loadings and high conversions. Nevertheless, the optimum amount of potassium promoter needs to be determined for each catalyst and FTS conditions.

Copper addition to Fe-based Fischer-Tropsch catalysts has been also extensively used because of its ability to improve the reducibility of Fe oxides. When copper oxide is reduced to metallic Cu, the crystallites formed provide H_2 dissociation sites, which in turn lead to reactive hydrogen species. Iglesia *et al. [44]* have shown that these species increase the rates of reduction and carburization of Fe precursors and of nucleation of reduced Fe-containing phases. These faster nucleation rates reflect a larger number of nuclei, which ultimately lead to higher active surface areas. These authors observed the same effect when potassium is added to catalyst. The previously reported decrease in FTS rates at low conversions with increasing K content was not observed in this case. In a doubly promoted sample with Cu and K, the effects are nearly additive, suggesting an almost independent effect of each promoter. A promoter surface density of 1 and 2 atoms nm^{-2} for Cu and K, respectively (normalized by the precursor oxide surface area) was found to yield Fe catalysts displaying the highest FTS rates.

Some attempts to improve the density of active sites and to decrease the selectivity to CO_2 have led to the replacement of Cu by Ru [44]. The addition of Ru instead of Cu increases significantly the hydrocarbon synthesis rate while decreasing slightly the selectivity to carbon dioxide. Similarly to Cu, the presence of Ru is reflected in higher reduction and carburization rates compared to the unpromoted samples. However, Ru forms smaller crystallites of reduced Fe species than Cu, which yields catalysts with higher FTS rates. Zr and Cr promoters increase the catalytic activity of Fe-based catalysts by increasing the number of active surface intermediates on the iron surface, leading thus to higher yields to hydrocarbon products [124].

Manganese has been reported to act as a chemical and structural promoter of Fe catalysts. The catalytic behaviour of mixed iron-manganese oxides is found to be influenced by the preparation technique and the structural properties of the catalytic precursors [108,125]. The preparation of these systems by the microemulsion technology results in an improved interaction between the two components, rendering more homogeneous samples than the traditional coprecipitation route and enhancing the Fe-Mn interaction [102]. Moreover, the catalytic performance of the samples depends on the Fe/Mn atomic ratio, being necessary a small amount of manganese (only 5 wt.%) to enhance the catalytic activity and improving the hydrocarbons formation rate. Furthermore, it is significant that Mn promotion leads to a higher olefin-to-paraffin ratio [103]. A very similar effect has been detected when using Ce as a promoter. In this case, the activity of the catalyst is enhanced when Fe-Ce interactions are developed, which can be achieved preparing Fe-Ce solid solutions by the coprecipitation technique. The specific identity of the Fe-Ce structure obtained depends on the relative concentration of Fe and Ce. Thus, Fe and Ce atoms are dissolved within the ceria or hematite structure, respectively [95]. Further studies have shown that, irrespective of preparation method, a solid solution is formed after treating the material at high temperature when a microscopic contact between Fe and Ce cations in the precursor is present [94]. This was evidenced by several characterization techniques, including X-ray diffraction, and Raman and Mössbauer spectroscopic techniques [94,95]. An important consequence of the formation of Fe-Ce mixed oxides is that the solid surface area is significantly increased. As a result, those catalysts displaying a higher degree of iron-cerium interactions show a better catalytic performance. The promotion of Fe catalysts with Ce can be directly related to the formation of Fe-O-Ce bridges in the calcined solids. Thus, it was proposed that the Ce promoter effect is consequence of Fe^0-Ce(III) ensembles arising from Fe-O-Ce bridges [94].

Precipitated iron catalysts typically contain a structural promoter to prevent total collapse of the highly porous precipitated iron oxide/hydroxide upon calcination and reduction [45]. Structural promoters with a good activity are silica, alumina and zinc oxide [44]. Another approach to obtain catalysts with attrition resistance in FTS is the use of Fe supported catalysts. In general, these materials display lower FTS rates than the unsupported catalysts. Nevertheless, the use of alternative preparation methodologies (microemulsion) has led to Fe/SiO_2 catalysts with higher catalytic rates (normalized per mass of Fe) than the unsupported reference [109].

In summary, a typical Fe-based catalyst for FTS consists of bulk Fe oxide promoted several elements. Copper is used to increase the reduction and carburization rates (forming thus a higher number of active sites); an alkaline metal, preferentially K, acts as electronic and textural promoter to improve the rate and selectivity towards large hydrocarbons and olefins. Finally, a structural promoter is required to provide a high surface area and to improve attrition resistance.

2.3. Activation of Fe-Based Catalysts

Co-based catalysts are typically activated in H_2 stream at 470-720 K to transform cobalt oxide into metallic Co atoms, which are the active sites in

FTS [66]. In contrast, the activation protocol is much more complicated for Fe-based catalysts. The activation of Fe catalysts can be carried out using CO [126,127], H_2 or H_2/CO mixtures [128,129]. During the pretreatment step, and even during the FTS reaction, Fe species develop several phase transformations. The identity of the Fe phase formed after the pretreatment step influences significantly the catalytic performance of the catalyst, especially at the early stages of the reaction. Nevertheless, the Fe phase initially formed evolves with time as the reaction proceeds.

During the activation step, the iron oxide precursor (usually hematite, α-Fe_2O_3) is transformed into magnetite (Fe_3O_4), irrespective of the activation gas used for the pretreatment. From this point, different iron phases can be formed depending on the activating atmosphere. When H_2 is used, Fe_3O_4 is converted into metallic Fe (α-Fe^0). In contrast, the use of CO or H_2/CO mixtures results in the formation of metallic Fe and different Fe carbides species, including metastable O-carbides (ε-Fe_2C, ε'-$Fe_{2.2}C$ and FeC_x,) and TP-carbides (χ-$Fe_{2.5}C$ and θ-Fe_3C) [130]. Fe carbides are formed when carbon atoms from CO dissociation are dissolved into the α-Fe^0 lattice. The different Fe carbides can be easily identified and characterized by X-ray diffraction techniques because of the changes in lattice parameters of the crystal; Mössbauer spectroscopy is also useful to detect different Fe carbides. In contrast, characterization techniques insensitive to the coordination of the atoms (as X-ray photoelectron spectroscopy, XPS) can detect the presence of Fe carbides, but cannot provide information about the specific type of Fe carbide. As mentioned above, Fe precursors treated in H_2 yield exclusively metallic Fe, but this structure evolves into Fe carbides during FTS reaction [131].

These general relationships between pretreatment atmospheres (H_2, CO or H_2/CO) and Fe phase formed are also influenced by the other factors. Thus, the identity of Fe structure formed during the activation step will also depend on the time of exposure to the reactant feed, the composition of this feed, the reactor system and the activation conditions (temperature and pressure) [132].

Although the exact identity of the active Fe phase (Fe^0, FeC_x or FeO_y) in FTS is still controversial [133-135], a correlation between the carbide content and the rate of hydrocarbons formation has been widely observed [136,137]. Moreover, several carbonaceous species are also formed on the catalyst surface during the activation process in CO or H_2/CO. These carbon species are of great importance in FTS catalysis because they are representative of the surface reaction intermediates. Therefore, the identification and the establishment of a structure–activity relationship between Fe phases and surface carbon species is a critical point to improve the FTS rate and tune the reaction selectivity.

Temperature-programmed surface reaction with H_2 (TPSR-H_2 or TPH) has been used to identify the Fe carbides and surface carbon species formed during pretreatment and/or FTS reaction. Bartholomew *et al.* [138,139] have identified the following species using this technique: C_α (atomic carbonaceous species resulting from CO dissociation), C_β (polymeric surface species with 2-3 carbon atoms), $C\gamma$ (iron carbides, mainly, θ-Fe_3C and χ-$Fe_{2.5}$), and C_δ (graphite-like species). The temperature required to hydrogenate these carbon species to methane (the only hydrogenation product obtained at ambient pressure) is indicative of the degree of their reactivity. Experimentally, it is found that the reactivity decreases in the following order: $C_\alpha > C_\beta > C\gamma > C_\delta$.

By combining *in situ* temperature programmed treatments using H_2 (TPH) and inert gases (TPD), *ex situ* chemical and structural characterization (XRD, Mössbauer spectroscopy, Raman spectroscopy) after passivation, and measurement of the catalytic activity, the identification of the surface carbonaceous and Fe phases (formed after different pretreatments and stages of the FTS reaction) has been accomplished [140]. The influence of the pretreatment atmosphere (CO, CO/Ar, H_2/CO or H_2) has been investigated and correlated with the presence and abundance of surface carbon and Fe species, as well as with the changes and evolution of such species during the FTS reaction, establishing thus a relationship between the nature of the Fe phase and surface carbon species and their abundance during the different kinetics episodes of the reaction. In this scenario, the effect of Ce and Mn addition on the identity of the Fe species has been extensively investigated by our group. Figure 2.3 summarizes the catalysts, reacting atmospheres and characterization techniques used in these studies.

As mentioned above, the temperature at which methane is formed in TPSR-H_2 experiments is indicative of the reactivity of the surface carbon species. Thus, methane formation from C_α species is observed at ~770 K when a pure iron oxide (hematite) solid is pretreated in CO. In contrast, these species are not detected when the pretreatment is carried out with H_2/CO or when catalytic promoters (Ce or Mn) are added. The temperature required to form methane from C_β species (polymeric surface species with 2-3 carbon atoms) varies between 870 and 950 K, depending again on the catalyst pretreatment and composition. Methane formation via hydrogenation of iron carbide species (θ-Fe_3C or χ-$Fe_{2.5}$) occurs at temperatures above 1000 K.

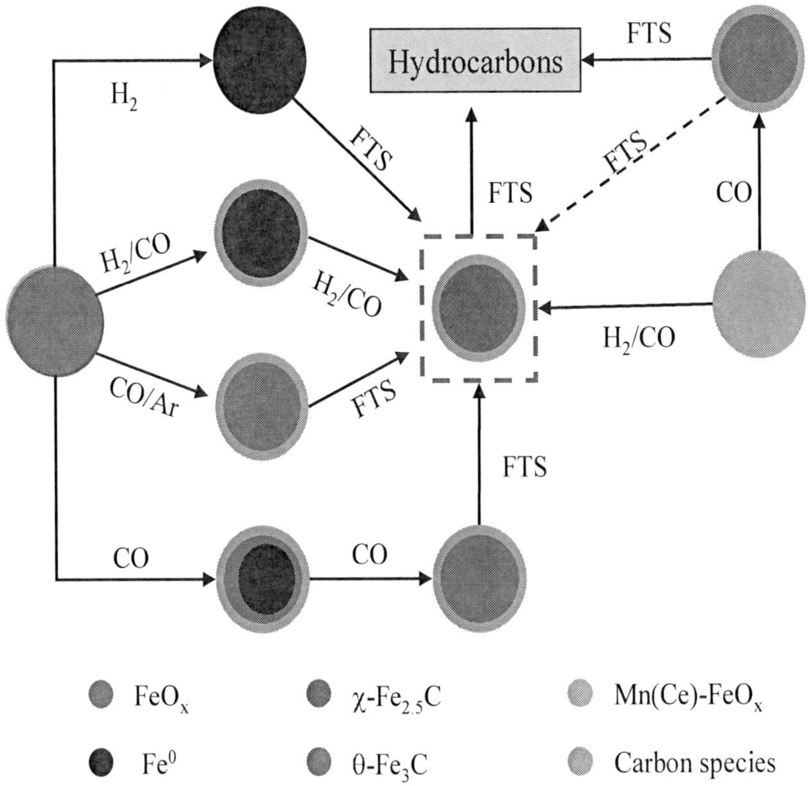

Figure 2.3. Transformation routes of iron oxide during pretreatment and FTS reaction.

A correlation between the surface carbonaceous species and the identity of the iron carbides formed after the pretreatment step has been found by combining TPSR-H_2 experiments with characterization data obtained from X-ray diffraction, and Raman and Mössbauer spectroscopy of the Fe samples after the activation step. Previously to the characterization experiments, the samples were passivated by flowing an 1 vol.% O_2/He mixture at room temperature for 1 h, according to the procedure reported previously in the literature [96]. The treatment of Fe catalysts with H_2/CO mixtures causes the transformation of Fe_2O_3 precursor into the Hägg carbide (χ-$Fe_{2.5}C$), evidenced by XRD and Mössbauer spectroscopy. TPSR-H_2 experiments of unpromoted and promoted Fe samples pretreated with H_2/CO revealed the exclusive formation of C_β species (polymeric surface species). On the contrary, the pretreatment of the unpromoted Fe_2O_3 precursor with CO originates predominantly cementite (θ-Fe_3C). Moreover, the TPSR-H_2 experiments show that both C_α and C_β species are formed over the cementite-type carbide. When Ce and Mn promoters are added to the Fe_2O_3 precursor, the treatment with CO renders the formation of cementite carbide (θ-Fe_3C) with C_α and C_β carbon species, although the relative concentration of C_β species is higher in this case. This clearly reveals that the catalyst composition and pretreatment procedure affects the nature of the surface species. Obviously, the pretreatment with H_2 form α-Fe^0 in both the unpromoted and promoted Fe samples.

Unpromoted and Ce- and Mn-promoted Fe catalyst were tested in the FTS reaction at 573 K and 1.01 MPa for ~170 h after the activation pretreatment with different reactive atmospheres (H_2, CO, CO/Ar and H_2/CO). As a general trend, the pretreatment with H_2/CO mixtures yields the most active catalysts, whereas the use of H_2 leads to catalysts with low activity. Those samples pretreated with CO (either pure or diluted with Ar) require Ce or Mn promotion to show relatively high reaction rates.

The CO conversion rate (normalized by the number of Fe atoms) of the unpromoted and Ce- and Mn-promoted samples is very similar when the pretreatment is performed in H_2/CO mixtures. In contrast, the activation in CO reveals significant differences in reaction rates, following the order: FeCe > FeMn > Fe. A common feature observed for all the catalysts is the increase of the catalytic activity after at certain time-on-stream period (*ca.* 15 h). This behaviour has been ascribed to a surface reconstruction of the catalyst [86]. Although the structure of the catalyst surface evolves during the reaction, the product selectivity does not change with the reaction time, indicating that more active sites are formed but without changing the catalytic properties.

By combining the catalytic activity data with the TPSR-H_2 experiments, it can be concluded that in general, those samples containing high concentrations of C_β species display high FTS rates. In addition, the unequivocal identification of the different type of Fe carbides by XRD and Mössbauer spectroscopy allowed the correlation of iron carbides and surface carbon species. Thus, it is found that C_β species are preferentially formed over the Hägg carbide (χ-$Fe_{2.5}C$). In the case of the Ce- and Mn-promoted samples activated in CO, although the cementite-type carbide (θ-Fe_3C) is the most abundant species, a higher amount of C_β species is stabilized compared with the unpromoted Fe sample.

Interestingly, the total amount of iron carbides formed during the pretreatment step is not as critical as its nature. This feature is clearly demonstrated from the experiments consisting in treating the catalyst with pure or diluted CO. When the catalyst is treated in pure CO, iron oxide hematite (α-Fe_2O_3) evolves predominantly to cementite (θ-Fe_3C). However, when the treatment is carried out in diluted carbon monoxide (5 vol.% CO/Ar), most of the iron remains in the form of oxides (hematite and magnetite) and only a small amount of cementite is found by XRD. Surprisingly, the treating in CO/Ar leads to more active catalysts in FTS, despite the fact that a higher amount of iron carbide species are formed when the treatment is carried out in pure CO. Mössbauer analysis after FTS reaction of the iron catalysts treated with diluted CO showed that the iron oxide present after the pretreatment has evolved to Hägg carbide (χ-$Fe_{2.5}C$). In the case of H_2 activated samples, Hägg carbide was also found after reaction using Mössbauer spectroscopy. This observation strongly suggests that irrespective of the gas used for the pretreatment, the Hägg carbide (χ-$Fe_{2.5}C$) is formed under the FTS reaction conditions. Therefore, Hägg-carbide seems to be the active phase responsible for the formation of hydrocarbons from H_2/CO mixtures. This is further supported by the fact that the catalytic activity of the samples pretreated with CO increased at some point after the beginning of the reaction [140]. Mössbauer spectroscopy analysis of the CO-activated samples after reaching the steady state in reaction reveals that the cementite (θ-Fe_3C) present after the pretreatment evolves partially to form Hägg carbide (χ-$Fe_{2.5}C$). However, the activity of these samples is not as high as the samples pretreated with H_2/CO because the extent of this transformation is limited. This result evidences a transformation of cementite into Hägg carbide during reaction concomitant with an increment of the catalytic activity.

The pretreatment with H_2 leads to less active catalysts compares to the use of H_2/CO mixtures, despite the fact that all the metallic iron atoms evolve to Hägg carbide (χ-$Fe_{2.5}C$) during reaction. The reason is that these bulk Fe samples sinter during the reduction in hydrogen. This causes the decrease of both the surface area and the catalytic activity compared to the samples activated in synthesis gas.

Regarding the Mn- and Ce-promoted samples, the situation is slightly different when the activation is carried out with CO. Although most of the iron is transformed into cementite during the pretreatment, the amount of carbonaceous C_β species detected by TPSR-H_2 is higher than in the unpromoted iron sample and comparable to the amount found in the samples activated in syngas (H_2/CO). This explains the higher activity of the promoted samples, even though the type of carbide found after treatment (cementite) is not the most suitable to stabilize the active carbonaceous intermediates (C_β).

In summary, when iron oxide (hematite) is pretreated with H_2, CO and H_2/CO, different iron phases are formed. Only metallic iron is found when the pretreatment is carried out with hydrogen. Iron carbide, mainly cementite (θ-

Fe_3C), is formed with CO (pure or diluted) pretreatments. On the other hand, Hägg carbide (χ-$Fe_{2.5}C$) is formed when the pretreatment is performed with syngas (H_2+CO). Surface carbonaceous species are stabilized on both types of iron carbides: C_α species are stabilized on cementite (θ-Fe_3C), whereas C_β species are stabilized on the Hägg carbide (χ-$Fe_{2.5}C$). When θ-Fe_3C species are present, the activity in the FTS is lower, although it can evolve in the reaction medium (H_2/CO at ~573 K) into the more active species, the Hägg carbide, where the C_β intermediates are stabilized. The extent of this transformation is not complete, and therefore, some cementite species remain on the catalyst surface, explaining thus the lower activity of the catalysts activated in CO. Mn and Ce promotion is only effective when the samples are activated with CO, favoring thus the stabilization of C_β species.

2.4. Dynamic Character of the Catalytically Active Fe Phases

The composition of the bulk and surface of Fe-based catalysts suffers severe changes during pretreatment steps and the FTS reaction. These modifications of the Fe phases are responsible for the typical different kinetic episodes that are observed experimentally. Hence, Schulz *et al.* [131] reported that the approach to the steady-state with Fe-based catalysts can be separated into several episodes of distinct kinetic regimes (induction periods) when the catalysts are activated with H_2. These episodes consist in carbon deposition and Fe carburization processes until the totality of the α-Fe (the main phase after H_2 pretreatment) is transformed into χ–$Fe_{2.5}C$. At this point, the steady-state is reached and the formation of the *true* FTS catalyst is supposed to occur. However, some authors have not found a correlation between the amount of bulk iron carbide and the activity of the catalyst [141]. They proposed that the bulk iron carbide serves merely as a support of the active surface species.

Goodwin *et al.* [142] investigated the differences between unpromoted and Mn- and K-promoted Fe catalysts concerning the induction period required to reach the steady-state in FTS. These authors determined the turnover frequency of the catalysts using steady-state isotopic transient kinetic analysis (SSITKA). The specific reaction rates were similar for all catalysts. The addition of Mn and/or K to the Fe catalyst leads to shorter induction periods and a higher number of active sites. However, the nature and the evolution of the active Fe phase were not reported in this study.

As already described, Bartholomew *et al.* [139] have identified and quantified different Fe carbide and carbonaceous species on the surface of unsupported and supported Fe catalysts by TPSR-H_2. A correlation between the coverage of atomic surface carbon species and the catalyst activity was found in both cases. Herranz *et al.* [103] have also identified different carbonaceous and Fe carbide species on Mn- and Ce-promoted Fe catalysts after different activation pretreatments by using the same TPSR technique. A straightforward correlation between the concentration of the Hägg carbide (χ-$Fe_{2.5}C$) species and the catalytic activity was found. Furthermore, the higher activity of the catalysts was directly related to a higher coverage of polymeric surface carbon species (C_β).

Hitherto, the actual nature of the Fe active phase(s), the surface species and their evolution in the different kinetic regimes has not been unequivocally determined with these studies. Unfortunately, these changes cannot be followed with *in situ* characterization techniques due to the technical complications derived from the stringent reaction conditions (high temperature and pressure). However, a successful experimental strategy to shed light on the nature of the Fe phase transformations occurring during FTS reaction has been recently developed [143]. The first step in this approach involves pre-treating the catalysts with a H_2/N_2 mixture (1:2 ratio) at 673 K to form α-Fe^0 that evolves during the reaction. In this way, the different kinetic episodes can be easily followed. The catalysts are tested in the FTS for different time-on-stream (TOS) values and the surface and bulk composition is studied using several *ex situ* characterization techniques (TPSR-H_2, N_2 adsorption-desorption isotherms, XRD and Mössbauer spectroscopy). This procedure requires the use of a suitable passivation protocol to avoid undesirable changes in the catalyst structure as a consequence of contact with ambient air when transferring the samples from the reactor to the characterization instruments. These studies have been conducted over both unpromoted (referred to as Fe) and two 5 at.% Ce-promoted Fe catalysts. The Ce-promoted samples were prepared by impregnation of either iron oxyhydroxide (referred to as FeCe-I) or iron oxide (referred to as FeCe-IC) [143]. It has been observed that the rate of hydrocarbon formation depends on the catalyst structure and composition. Moreover, the reaction rate is also a function of time-on-stream. The unpromoted Fe catalyst displays a higher initial FTS rate compared to the promoted catalysts. However, the rate decreases monotonically as the reaction occurs. An inflection point is reached at a certain point (*ca.* 60 h on stream) and then the rate increases. The initial reaction rates of the Ce-promoted samples are lower compared to the unpromoted catalyst. In fact, the FeCe-IC sample presents a lower reaction rate in the whole range of TOS values because Fe-Ce interactions are not fully developed over this sample and that the addition of Ce actually results in a decreasing number of iron active sites. On the other hand, the catalytic performance of the sample FeCe-I, over which a chemical Fe-O-Ce interaction is effectively developed [95], exceeds the reaction rate of the unpromoted catalyst after *ca.* 15 h on stream, despite the fact that the initial activity of this sample is the lowest among all the catalysts studied here. The BET surface area of the different catalysts noticeably changes in parallel with the hydrocarbon formation rate as the reaction proceeds. Figure 2.4 collects the BET and hydrocarbon formation rate *vs.* time on stream for those catalysts. We have not found a straightforward correlation between the BET surface area and catalytic activity at the early stages of the reaction. However, this is detected as the reaction proceeds. Therefore, the reaction rate is determined by a combination of several factors, not only by catalyst surface area. This is clearly seen in the case of the FeCe-I sample, which records the highest surface area among all the materials yet the lowest reaction rate. Other features should therefore be called upon to explain both the initial activity of the catalysts and the evolution of the reaction rate as the FTS proceeds.

X-ray diffraction patterns showed that the carburization process of the Ce-containing samples is retarded compared to the unpromoted Fe catalyst. For the Ce-promoted solids, the presence of metallic iron (α-Fe), magnetite (Fe_3O_4) and iron carbides (χ-$Fe_{2.5}C$ and θ-Fe_3C) during the early stages of the reaction was evidenced. By contrast, magnetite and iron carbides were observed for the unpromoted samples. Thus, the higher activity of the Ce-free sample during the early stages of the FTS is explained in terms of its higher carburization degree as compared to the Ce-containing samples. The XRD analysis of this unpromoted catalyst used in the FTS for 40 h on stream indicates that the Hägg carbide becomes the predominant Fe phase at this point, being thus responsible for the increasing rate as the reaction proceeds. However, the situation for the Ce-promoted catalysts is slightly different. In this case, the rate of hydrocarbon formation increases even before the θ-Fe_3C phase evolves to χ-$Fe_{2.5}C$, although χ-$Fe_{2.5}C$ is the predominant carbide phase on the FeCe catalysts, as shown in Figure 2.5.

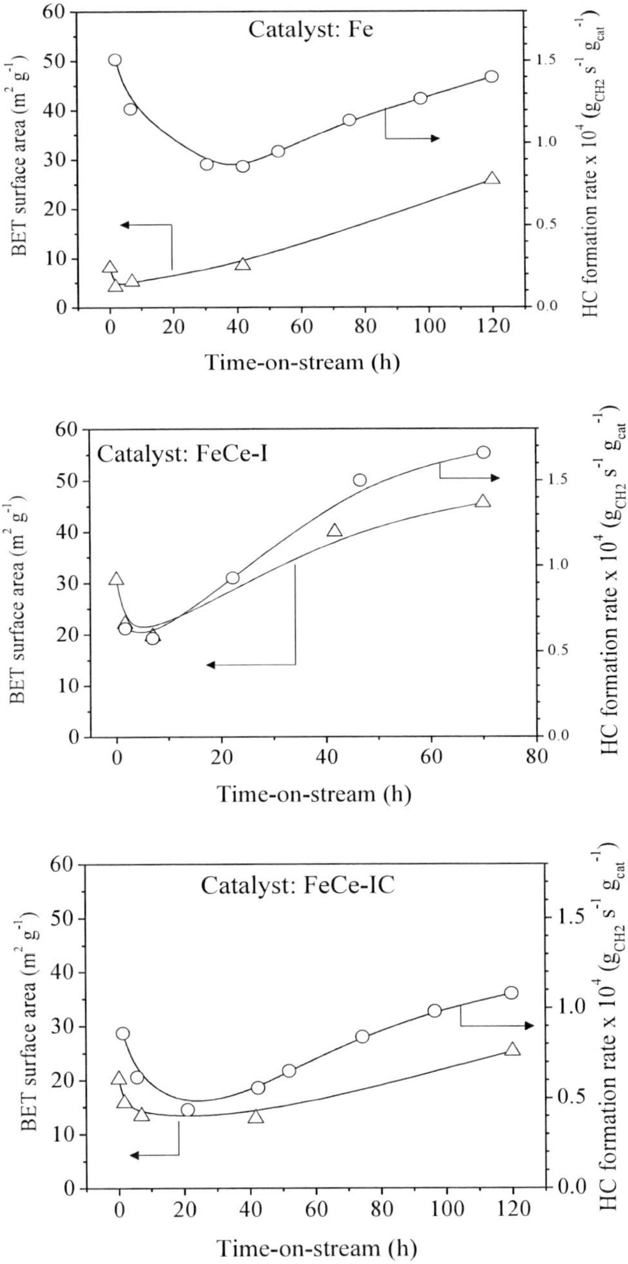

Figure 2.4. Evolution of the BET surface area and hydrocarbon (HC) formation rate during time-on-stream observed with the unpromoted and Ce-promoted Fe samples.

These observations have been further corroborated by Mössbauer spectroscopy. Both characterization techniques, Mössbauer and XRD, show a lower carburization extent of the Ce-promoted samples. The amount of the Hägg carbide observed in the unpromoted Fe catalyst increases at the expense of the cementite as the FTS proceeds, and eventually, the χ-$Fe_{2.5}C$ structure is the only carbide phase detected. By contrast, both iron carbide phases (cementite and Hägg) are always detected in the Ce-containing samples, although a continuous transformation of cementite into the Hägg carbide along the FTS reaction is observed. The Ce-promoted catalysts develop more active materials than the unpromoted sample once the catalytic precursor is totally transformed into iron carbides (either as cementite or Hägg carbide). It should also be noted that TPSR-H_2 characterization found a direct correlation between the concentration of C_β on the catalyst surface and the catalytic activity for all catalysts.

This work evidences the reconstruction process that Fe-based catalysts undergo during the FTS. As it was pointed out above, a direct correlation between the concentration of C_β species (partially polymerized carbon species) on the catalyst surface and the catalytic activity of all the samples for time-on-stream values higher than 7 h was found. In addition, the increase of the concentration of this type of species was parallel with the increase of the surface area of the solids. In the case of the Ce-free iron sample, the increase of the catalytic activity was also accompanied by an increase of the concentration of χ-$Fe_{2.5}C$ species and by a decrease of the θ-Fe_3C concentration. According to the XRD and Mössbauer spectroscopy results, this evolution was observed continuously during the reaction. Finally, the θ-Fe_3C phase could not be detected after 120 h on stream. The Ce-free iron catalyst underwent the increase of the catalytic activity after 40 h on stream, which coincides with the total transformation of the θ-Fe_3C phase into χ-$Fe_{2.5}C$ and with the stabilization of the graphitic deposits concentration on the surface. In summary, a good correlation between χ-$Fe_{2.5}C$ and FT activity was found for the Ce-free iron catalyst.

In the case of the Ce-loaded iron catalysts, the transformation of θ-Fe_3C species into χ-$Fe_{2.5}C$ is also observed, although the θ-Fe_3C species did not evolved totally. Considering that the FeCe catalysts displayed a shorter induction process, Ce addition would be responsible of the formation of C_β species, independently of the graphitic deposits formation. The effect of Ce addition could be also related with the increase of the surface area. Despite the fact that the θ-Fe_3C did not evolve totally into χ-$Fe_{2.5}C$ in the case of the Ce-loaded iron catalysts, Ce addition originated a higher number of active sites, and consequently, the effect of graphitic deposits is less significant in absolute terms.

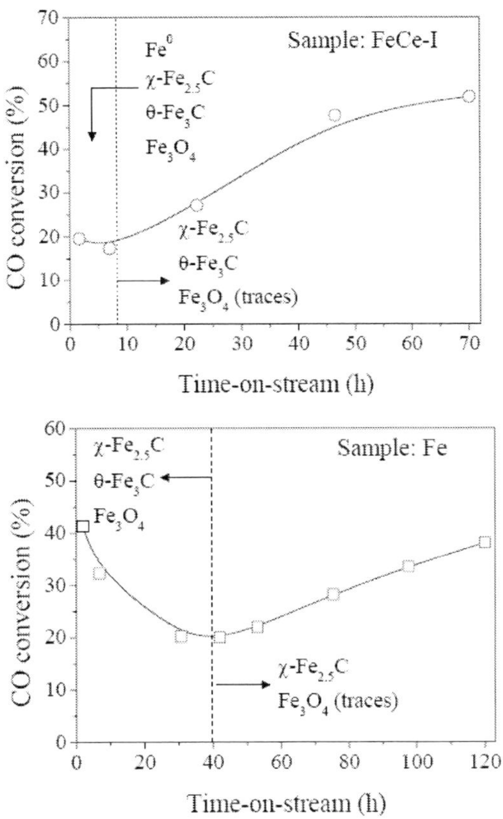

Figure 2.5. Relationship between the Fe phases detected (XRD and Mössbauer spectroscopy) and the FTS kinetic episodes with FeCe-I and Fe catalysts.

With regard to the promoter effects, it was speculated that Ce-promoted catalysts show higher reaction rates because of the formation on the catalyst surface of tilted CO species, which are easier to dissociate. Ce (III) species are needed in the framework of the catalysts in order to form such tilted CO species. This is supported by the X-ray photoelectron spectroscopy (XPS) results obtained after treating the samples in H_2/CO mixtures, showing the presence of these Ce(III) species [95] .

In summary, the nature of the surface species and bulk composition of iron-based catalyst is dynamic and depends on the time-on-stream. Several iron phases, including iron oxides, metallic iron and different iron carbides are observed initially but the increase of the catalytic activity of iron based catalysts is directly related with the evolution of iron carbide species to Hägg carbide species (χ-$Fe_{2.5}C$) and the consequently formation of C_β species on the catalyst surface. The addition of certain additives as Ce to iron oxide FT catalysts increases the catalytic activity in terms of a faster activation and a higher catalytic activity level in the steady-state. This is due to the stabilization of higher amounts of active reaction intermediates (C_β).

2.5. Fischer-Tropsch Synthesis with CO_2-Containing Syngas

The increasing global energy demand during the past decades has caused an enormous raise of the man-made CO_2 emissions. Many efforts have been devoted in order to reduce such CO_2 emissions as a consequence of the Third Conference of the Parties (COP-3) in Kyoto in 1997 [144]. Furthermore, the improvement of the efficiency of energy conversion or utilization processes strategies requires other secondary approaches, including capture, storage, and fixation of the carbon dioxide produced [145]. In this context, the production of clean fuels via CO_2 hydrogenation appears as a very interesting and attractive option from an industrial point of view.

Hydrogenation of carbon dioxide has been traditionally carried out with catalysts that have been demonstrated to be active and selective for the FTS reaction, mainly because CO_2 may be a significant component in the synthesis gas fed to the FTS plants [137]. Thus, CO_2 is an important component in the synthesis gas obtained via natural gas partial oxidation or steam reforming, in the syngas obtained from biomass gasification and also in the feed to the industrial FTS reactors that recycle unconverted reactants. Moreover, CO_2 may be present in the synthesis gas produced through the carbon dioxide reaction with methane, a route of great interest in recent years because it constitutes a promising alternative to utilize the CO_2 originated in various anthropogenic processes. The removal of CO_2 from H_2/CO mixtures is a very expensive process. Therefore, the development of catalysts active in CO_2 hydrogenation to hydrocarbons is a key step to increase the feasibility of FTS overall processes [69,146].

K-promoted Fe-based catalysts have exhibited the most promising results in the CO_2 hydrogenation to form long-chain hydrocarbons [147,148], although promotion with other elements (Cr, Mo, Mn and Zn) have been also addressed [102,149,150]. These studies have indicated that the reaction proceeds through the reverse water-gas shift (RWGS) reaction ($CO_2 + H_2 \rightarrow CO + H_2O$). According to this mechanism, CO_2 is first converted into CO, which is subsequently hydrogenated to form different hydrocarbons. Consequently, the direct CO_2 hydrogenation to hydrocarbons is irrelevant, and the reaction always takes place through the formation of CO intermediates [102,147]. Therefore, an important requirement for the CO_2 hydrogenation catalysts in order to be active is that they must be able to catalyze the RWGS reaction. Iron-based materials are active catalysts for both RWGS and FTS reactions, and as a result, they can be used to obtain hydrocarbons from H_2/CO_2 mixtures.

Fe catalysts attain the steady state activity in FTS reaction through *in situ* activation as a consequence of the catalyst reconstruction that involves the formation of different types of Fe carbides [137,143]. Riedel *et al.* [148] showed that the reconstruction of Fe catalyst is similar when using either H_2/CO or H_2/CO_2 mixtures, although the kinetic regimes in CO_2 hydrogenation are up to 10 times slower than in H_2/CO reaction because a much lower concentration of CO is available to form the Fe carbides. It is also remarkable that the product selectivity was found to be similar at steady-state in both CO and CO_2 hydrogenation reactions. This suggests common active sites for both reactions at the steady-state.

The hydrogenation of CO_2 with unpromoted and Ce-promoted Fe catalysts has been studied recently [151]. Cerium is known to promote both the WGS and RWGS reactions [146], which is a critical requirement in order to activate the carbon dioxide molecules. Ce addition to Fe-based catalysts leads to shorter induction periods in the formation of hydrocarbons from CO_2 hydrogenation, which is due to a higher carburization rate of the Ce-promoted Fe catalyst. In contrast, a similar product distribution is found with the unpromoted and Ce-promoted catalysts.

The hydrogenation of CO_2 to hydrocarbons is limited by the approach to the equilibrium at high conversions. This limitation reduces the CO available to form Fe carbides, which is reflected in longer induction periods. This can be partially avoided using higher operation temperatures. In this way, a higher concentration of CO is formed, but possesses negative effects on the selectivity of CO hydrogenation.

A very recent thermodynamic study, based in improving the energy used in the FT process, reveals that by using CO_2 instead of CO the overall energy efficiency of the hydrocarbon synthesis process can be increased [152]. The conventional FT process is inefficient because it requires a large amount of work to be put into the gasification step and emits too much work during the FT synthesis. By producing CO_2 rich syngas, the amount of work required by the gasifier diminished, hence the energetic balance can be improve, especially since making hydrocarbons from CO_2 is less exothermic[153]. Needless to say, that this approach would be only feasible if carbon-free H_2 is available and *"in-situ"* reverse water gas shift reaction is needed to produce CO in the FTS reactor.

In conclusion, although more research about process conditions and catalyst composition is needed to obtain high yields to hydrocarbons in CO_2 hydrogenation reactions, obtaining hydrocarbons from CO_2 rich syngas could have significant advantages over the traditional process.

Chapter 3

ECONOMICS, FUTURE PROSPECTS AND CONCLUDING REMARKS

The production of hydrocarbons via FTS is a sustainable route to liquid transportation fuels. The BTL-process is one of the more attractive options because the possibility of using the large reserves of biomass. Fe-based catalysts are especially suitable because of the low cost and the ability of converting syngas mixtures with low H_2/CO ratios to different hydrocarbons. Nevertheless, adequate catalyst preparation methods and addition of catalytic and structural promoters is required to maximize the yield to hydrocarbons and to tune the product distribution.

Fuels and chemical derived from FTS compete with traditional oil, and therefore, the economic feasibility of producing these fuels and chemicals via FTS would be ultimately dominated by the price of oil. Moreover, different factors, such us reducing liquid fuel dependence, obtaining liquid fuels from coal or strained natural gas, and environmental benefits derived from the BTL process, could play a major role in obtaining tax exemptions for synthetic FTS fuels.

Currently, FTS fuels cannot compete with traditional fossil fuels without these incentives. Although somehow scarce, there are open public studies dealing with the feasibility of obtaining liquid fuels from FTS via biomass gasification [14]. These documents coincide in that biomass gasification is not yet a mature technology and improvements in the gasification step, mainly by using pressurized gasifiers, should make FTS from biomass more competitive. Two independent studies reached similar conclusions [154,155]. Thus, the price of FT-liquid fuels would be around US$ 16/GJ (9.1-16.7 €$_{2002}$/GJ) in the short term. In 2002, oil reference price was roughly 2.6-7.0 €$_{2002}$/GJ. Since then, oil price have significantly increased up to US$$_{2008}$150/barrel (US$$_{2008}$ 24.5/GJ) in the first part of 2008, although oil prices plummet from $150 to $40 per barrel in the second part of that year. Obviously, the FTS becomes competitive with traditional oil in the price scenario of years 2007 or early 2008. However, due to the large investments required to construct and operate a FTS plant, volatile market in oil price is an important drawback. Total capital investment for a FTS plant depends on various factors, including the total capacity or syngas source. In this sense, the investment needed for a FTS plant using natural gas as synthesis gas source is lower than that using biomass as a carbon source. This accounts for the pre-treatment and more complex gasification steps, as well as for the cleaning and water-gas shift units. The capital investment needed to construct a biomass-derived FTS operating with current technology is calculated to be around 286 M€ for a 400 $MW_{th,HHV}$ (thermal watts, high-heating value) input plant. FT diesel production price at such plant would be around 16.1 €/GJ. A major influence on the diesel price is the scale. Thus, FT diesel price could drop to *ca.* 14 €/GJ in a 2000 MW_{th} input plant.

There exist some strategies to the economics of the FTS plants. Serious attempts to do this should focus first and foremost in decreasing the price of the pre-treatment, gasification, cleaning and conditioning (shift) units. This part of the process accounts for *ca.* 75 % of the total capital investment of the FTS process. It is foreseen that in the long term, further drops of biomass price (2€/GJ) along with technological developments (oxygen production, catalytic tar cracking) could reduce capital costs around 15 % or 9 % in the FTS cost. For instance, developing more selective catalyst to the diesel fraction could avoid the upgrading unit.

In summary, the economic feasibility of GTL, CTL and/or BTL plants depends significantly on the production scale. Thus, large scale plants are required to produce synthetic fuels competitive with those obtained from crude oil.

REFERENCES

[1] Huber, G. W.; Iborra, S.; Corma, A. *Chem. Rev.* 2006, *106*, 4044-4098.

[2] *Energy Information Administration U.S.A.,* 2008, *International Energy Outlook*, http://www.eia.doe.gov/oiaf/ieo/pdf/0484(2008).pdf.

[3] *Energy Information Administration U.S.A.,* 2009, *Annual Energy Outlook*, http://www.eia.doe.gov/oiaf/aeo/pdf/earlyrelease.pdf.

[4] Sabatier, P.; Senderens, J. B. *J. Chem. Soc., Faraday Transac.* 1902, *82*, 333-337.

[5] Fischer, F.; Tropsch, H. *Brennst Chemie* 1923, *4*, 276.

[6] *http://www.choren.com/en/biomass_to_energy/* 2009,

[7] Balat, M.; Balat, M.; Kirtay, E.; Balat, H. *Energy Conversion Management* 2009, *50*, 3158-3168.

[8] Hallett, W. L. H.; Clark, N. A. *Fuel* 2006, *85*, 532-544.

[9] Maniatis, K. "Progress in biomass gasification: an Overview," European Commission, 2001.

[10] Yung, M. M.; Jablonski, W. S.; Magrini-Bair, K. A. *Energy Fuels* 2009, *23*, 1874-1887.

[11] Wilhelm, D. J.; Simbeck, D. R.; Karp, A. D.; Dickenson, R. L. *Fuel Procces. Technol.* 2001, *71*, 139-148.

[12] Steynberg, A. P.; Nel, H. G. *Fuel* 2004, *83*, 765-770.

[13] Navarro, R. M.; Peña, M. A.; Fierro, J. L. G. *Chem. Rev.* 2007, *107*, 3952-3991.

[14] Tijmensen, M. J. A.; Faaij, A. P. C.; Hamelinck, C. N.; van Hardeveld, M. R. M. *Biomass Bioenergy* 2002, *23*, 129-152.

[15] Bergman, P. C. A.; van Paasen, S. V. B.; Boerrigter, H. *Pyrolysis and Gasification of Biomass and Waste Meeting* 2002,

[16] Iglesia, E.; Reyes, S. C.; Madon, R. J.; Soled, S. L. *Adv. Catal.* 1993, *39*, 221-302.

[17] Khodakov, A. Y.; Chu, W.; Fongarland, P. *Chem. Rev.* 2007, *107*, 1692-1744.

[18] van der Laan, G. P.; Beenackers, A. A. C. M. *Catal. Rev. Sci. Eng.* 1999, *41*, 255-318.

[19] van Steen, E.; Claeys, M. *Chem. Eng. Technol.* 2008, *31*, 655-666.

[20] Iglesia, E. *Appl. Catal. A: Gen.* 1997, *161*, 59-78.

[21] Bezemer, G. L.; Bitter, J. H.; Kuipers, H. P. C. E.; Oosterbeek, H.; Holewijn, J. E.; Xu, X.; Kapteijn, F.; van Dillen, A. J.; de Jong, K. P. *J. Am. Chem. Soc.* 2006, *126*, 3956-3964.

[22] Jacobs, G.; Das, T. K.; Zhang, Y.; Li, J.; Racoillet, G.; Davis, B. H. *Appl. Catal. A: Gen.* 2002, *233*, 263-281.

[23] Rioux, R. M.; Song, H.; Hoefelmeyer, J. D.; Yang, P.; Somorjai, G. A. *J. Phys. Chem. B* 2005, *109*, 2192-2202.

[24] Zheng, N.; Stucky, G. D. *J. Am. Chem. Soc.* 2006, *128*, 14278-14280.

[25] Martinez, A.; Prieto, G. *J. Catal.* 2007, *245*, 470-476.

[26] Sun, S.; Fujimoto, K.; Yoneyama, Y.; Tsubaki, N. *Fuel* 2002, *81*, 1583-1591.

[27] Moradi, G. R.; Basir, M. M.; Taeb, A.; Kiennemann, A. *Catal. Commun.* 2003, *4*, 27-32.

[28] Morales, F.; de Smit, E.; de Groot, F. M. F.; Visser, T.; Weckhuysen, B. M. *J. Catal.* 2007, *246*, 91-99.

[29] Krishnamoorthy, S.; Tu, M.; Ojeda, M. P.; Pinna, D.; Iglesia, E. *J. Catal.* 2002, *211*, 422-433.

[30] Storsaeter, S.; Borg, O.; Blekkan, E. A.; Holmen, A. *J. Catal.* 2005, *231*, 405-419.

[31] Kiss, G.; Kliewer, C. E.; DeMartin, G. J.; Culross, C. C.; Baumgartner, J. E. *J. Catal.* 2003, *217*, 127-140.

[32] Ojeda, M.; Pérez-Alonso, F. J.; Terreros, P.; Rojas, S.; Herranz, T.; López Granados, M.; Fierro, J. L. G. *Langmuir* 2006, *22*, 3131-3137.

[33] Bazin, D.; Kovacs, I.; Guczi, L.; Parents, P.; Laffon, C.; De Groot, F. M. F.; Ducreaux, O.; Lynch, J. *J. Catal.* 2000, *189*, 456-462.

[34] Salmeron, M.; Schlögl, R. *Surf. Sci. Rep.* 2008, *63*, 169-199.

[35] Morales, F.; de Groot, F. M. F.; Glatzel, P.; Kleimenov, E.; Bluhm, H.; Hävecker, M.; Knop-Gericke, A.; Weckhuysen, B. M. *J. Phys. Chem. B* 2004, *108*, 16201-16207.

[36] Herranz, T.; Deng, X.; Cabot, A.; Guo, J.; Salmeron, M. *J. Phys. Chem. B* 2009, *113*, 10721-10727.

[37] Yu, Z.; Borg, O.; Chen, D.; Enger, B. C.; Froseth, V.; Rytter, E.; Wigum, H.; Holmen, A. *Catal. Lett.* 2006, *109*, 43-47.

[38] Puntes, V. F.; Krishnan, K. M.; Alivisatos, A. P. *Science* 2001, *291*, 2115-2117.

[39] Beitel, G. A.; Laskov, A.; Oosterbeek, H.; Kuipers, E. W. *J. Phys. Chem.* 1996, *100*, 12494-12502.

[40] van Steen, E.; Claeys, M.; Dry, M. E.; van der Loosdrecht, J.; viljoen, E. L.; Visagie, J. L. *J. Phys. Chem. B* 2005, *109*, 3575-3577.

[41] Saib, A. M.; Borgna, A.; van der Loosdrecht, J.; van Berge, P. J.; Niemantsverdriet, J. M. *J. Phys. Chem. B* 2006, *110*, 8657-8664.

[42] Mitsui, T.; Rose, M. K.; Fomin, E.; Ogletree, D. F.; Salmerón, M. *Nature* 2003, *422*, 705-707.

[43] Inderwildi, O. R.; Jenkins, S. J.; King, D. A. *J. Phys. Chem. C* 2008, *112*, 1305-1307.

[44] Li, S.; Li, A.; Krishnamoorthy, S.; Iglesia, E. *Catal. Lett.* 2002, *77*, 197-205.

[45] Bukur, D. B.; Mukesh, D.; Patel, S. A. *Ind. Eng. Chem. Res.* 1990, *29*, 194-204.

[46] Koizumi, N.; Murai, K.; Ozaki, T.; Yamada, M. *Catal. Today* 2004, *89*, 465-478.

[47] Hindermann, J. P.; Hutchings, G. J.; Kiennemann, A. *Catal. Rev. Sci. Eng.* 1993, *35*, 1-127.

[48] Abrevaya, H.; Cohn, M. J.; Targos, W. M.; Robota, H. J. *Catal. Lett.* 1988, *7*, 183-195.

[49] Nurunnabi, M.; Murata, K.; Okabe, K.; Inaba, M.; Takahara, I. *Appl. Catal. A: Gen.* 2008, *340*, 203-211.

[50] Kang, J.; Zhang, S.; Zhang, Q.; Wang, Y. *Angew. Chem. Int. Ed.* 2009, *48*, 2565-2568.

[51] Tauster, S. J.; Fung, S. C.; Garten, R. L. *J. Am. Chem. Soc.* 1978, *100*, 170-175.

[52] Tauster, S. J.; Fung, S. C. *J. Catal.* 1978, *55*, 29-35.

[53] Krishna, K. R.; Bell, A. T. *J. Catal.* 1991, *130*, 597-610.

[54] Dry, M. E. *Appl. Catal. A: Gen.* 1996, *138*, 319-344.

[55] Murata, K.; Okabe, K.; Inaba, M.; Takahara, I.; Liu, Y. *Catal. Commun.* 2009, *128*, 343-348.

[56] Li, J.; Jacobs, G.; Zhang, Y.; Das, T. K.; Davis, B. H. *Appl. Catal. A: Gen.* 2002, *223*, 195-203.

[57] Ott, G. L.; Fleisch, T.; Delgass, W. N. *J. Catal.* 1979, *60*, 394-403.

[58] Adesina, A. A. *Appl. Catal. A: Gen.* 1996, *138*, 345-367.

[59] Bianchi, D.; Tau, L. M.; Borcar, S.; Bennett, C. O. *J. Catal.* 1983, *84*, 358-374.

[60] Biloen, P.; Helle, J. N.; Sachtler, W. M. H. *J. Catal.* 1979, *58*, 95-107.

[61] Biloen, P.; Helle, J. N.; van der Berg, F. G. A.; Sachtler, W. M. H. *J. Catal.* 1983, *81*, 450-463.

[62] Kobori, Y.; Yamasaki, H.; Naito, S.; Onishi, T.; Tamaru, K. *J. Chem. Soc., Faraday Transac.* 1982, *78*, 1473-1490.

[63] Ojeda, M.; Nabar, R.; Nilekar, A. U.; Ishikawa, A.; Mavrikakis, M.; Iglesia, E. *in preparation* 2010.

[64] Storch, H. H.; Golumbic, N.; Anderson, R. B., *The Fischer-Tropsch and related syntheses*; Wiley: New York, 1951.

[65] Storsaeter, S.; Chen, D.; Holmen, A. *Surf. Sci.* 2006, *600*, 2051-2063.

[66] Anderson, R. B.; Emmet, P. H., *The Fischer-Tropsch synthesis*; Academic Press Inc., 1956.

[67] van Dijk, H. A. J., Technische Universiteit Eindhoven, 2001.

[68] Pichler, H.; Schulz, H. *Brennst Chemie* 1967, *3*, 78.

[69] Dry, M. E. *J. Chem. Technol. Biotechnol.* 2001, *77*, 43-50.

[70] Madon, R. J.; Iglesia, E. *J. Catal.* 1993, *139*, 576-590.

[71] Falbe, J., *New syntheses with carbon monoxide*; Springer-Verlag: Berlin, 1980.

[72] Guettel, R.; Kunz, U.; Turek, T. *Chem. Eng. Technol.* 2008, *31*, 746-754.

[73] Dry, M. E.; Anderson, R. B. in: Boudart, M. (Ed.), Springer, Berlin, 1981, pp.

[74] Sethuraman, R.; Bakhsi, N. N.; Katikakeni, S. P.; Idem, R. O. *Fuel Process. Technol.* 2001, *73*, 197.

[75] Martinez, A.; Prieto, G. *Topics Catal.* 2009, *52*, 75.

[76] Shamsi, A.; Rao, V. U. S.; Gormley, R. J.; Obermeyer, R. T.; Schehl, R. R.; Stencel, J. M. *Appl. Catal.* 1986, *27*, 55.

[77] Egiebor, N. O.; Deckwer, H. D.; Ralek, M. *Appl. Catal.* 1989, *55*, 47.

[78] Baerns, M.; Guan, N.; Körting, E.; Lindner, U.; Lohrengel, M.; Papp, H. *Int. J. Energy Res.* 1994, *18*, 197.

[79] Rao, V. U. S.; Gormley, R. J. *Hydrocarb. Process.* 1980, *59*, 139.

[80] Hammer, H.; Joisten, M.; Lüngen, S.; Winkler, D. *Int. J. Energy Res.* 1994, *18*, 223.

[81] Bessell, S. *Appl. Catal. A: Gen.* 1995, *126*, 235.

[82] Martinez, A.; Valencia, S.; Murciano, R.; Cerqueira, H. S.; Costa, A. F.; Aguiar, E. F. S. *Appl. Catal. A: Gen.* 2008, *346*, 117.

[83] Martinez, A.; Rollán, J.; Arribas, M. A.; Cerqueira, H. S.; Costa, A. F.; Aguiar, E. F. S. *J. Catal.* 2007, *249*, 162.

[84] Jothimurugesan, K.; Gangwal, S. K. *Ind. Eng. Chem. Res.* 1998, *37*, 1181.

[85] Bessell, S. *Appl. Catal. A: Gen.* 1993, *96*, 253.

[86] Ngamcharussrivichai, C.; Imyim, A.; Li, X.; Fujimoto, K. *Ind. Eng. Chem. Res.* 2007, *46*, 6883.

[87] Botes, F. G.; Böhringer, W. *Appl. Catal. A: Gen.* 2004, *267*, 217.

[88] Udaya, V.; Rao, S.; Gormley, R. J. *Catal. Today* 1990, *6*, 207.

[89] Bao, J.; He, J.; Zhang, Y.; Yoneyama, Y.; Tsubaki, N. *Angew. Chem. Int. Ed.* 2008, *47*, 353.

[90] Lia, X.; Heb, J.; Menga, M.; Yoneyama, Y.; Tsubaki, N. *J. Catal.* 2009, *265*, 26.

[91] Cornell, R. M.; Schwertmann, U., *The iron oxides*; VCH Publishers: Weinheim, 2003.

[92] Ziolo, R. F.; Ginnelis, E. P.; Weinstein, B. A.; O'Horo, M. P.; Ganguly, B. N.; Mehrotra, V.; Russell, M. W.; Huffman, D. R. *Science* 1992, *257*, 219-223.

[93] Li, S.; Krishnamoorthy, S.; Li, A.; Meitzner, G. D.; Iglesia, E. *J. Catal.* 2002, *206*, 202-217.

[94] Pérez-Alonso, F. J.; López Granados, M.; Ojeda, M.; Herranz, T.; Rojas, S.; Terreros, P.; Fierro, J. L. G.; Gracia, M.; Gancedo, J. R. *J. Phys. Chem. B* 2006, *110*, 23870-23880.

[95] Pérez-Alonso, F. J.; López Granados, M.; Ojeda, M.; Terreros, P.; Rojas, S.; Herranz, T.; Fierro, J. L. G. *Chem. Mater.* 2005, *17*, 2329-2339.

[96] López-Quintela, M. A. *Curr. Opi. Colloid Interface Sci.* 2003, *8*, 137-144.

[97] Schwuger, J.; Stickdorn, K. *Chem. Rev.* 1995, *95*, 849-864.

[98] Eriksson, S.; Nylén, U.; Rojas, S.; Boutonnet, M. *Appl. Catal. A: Gen.* 2004, *265*, 207-219.

[99] Kishida, M.; Umakoshi, K.; Ishiyama, J.; Nagata, H.; Wakabayashi, K. *Catal. Today* 1996, *29*, 355-359.

[100] Capek, I. *Advances in Colloid and Interface Science* 2004, *110*, 49-74.

[101] Rojas, S.; Eriksson, S.; Boutonnet, M.; Royal Society of Chemistry, 2004.

[102] Herranz, T.; Rojas, S.; Ojeda, M.; Pérez-Alonso, F. J.; Terreros, P.; Pirota, K.; Fierro, J. L. G. *Chem. Mater.* 2006, *18*, 2364-2375.

[103] Herranz, T.; Rojas, S.; Pérez-Alonso, F. J.; Ojeda, M.; Terreros, P.; Fierro, J. L. G. *Appl. Catal. A: Gen.* 2006, *311*, 66-75.

[104] Jager, B.; Espinoza, R. *Catal. Today* 1995, *23*, 17-28.
[105] O'Brien, R. J.; Xu, L.; Bao, S.; Raje, A.; Davis, B. H. *Appl. Catal. A: Gen.* 2000, *196*, 173-178.
[106] Pham, H. N.; Nowicki, L.; Xu, J.; Datye, A. K.; Bukur, D. B.; Bartholomew, C. H. *Ind. Eng. Chem. Res.* 2003, *42*, 4001-4008.
[107] Chen, K.; Yan, Q. *Appl. Catal. A: Gen.* 1997, *158*, 215-223.
[108] Leith, I. R.; Howden, M. G. *Appl. Catal.* 1988, *37*, 75-92.
[109] Herranz, T.; Rojas, S.; Pérez-Alonso, F. J.; Ojeda, M.; Terreros, P.; Fierro, J. L. G. *Appl. Catal. A: Gen.* 2006, *308*, 19-30.
[110] Rockenberg, J.; Scher, E. C.; Alivisatos, A. P. *J. Am. Chem. Soc.* 1999, *121*, 11595-11596.
[111] Hyeon, T.; Lee, S. S.; Park, J.; Chung, Y.; Na, H. B. *J. Am. Chem. Soc.* 2001, *123*, 12798-12801.
[112] Sun, S.; Zeng, H. *J. Am. Chem. Soc.* 2002, *124*, 8204-8205.
[113] Sun, S.; Zeng, H.; Robinson, D. B.; Raoux, S.; Rice, P. M.; Wang, S. X.; Li, G. *J. Am. Chem. Soc.* 2004, *126*, 273-279.
[114] Kumar, R. V.; Koltypin, Y.; Xu, X. N.; Yeshurun, Y.; Gedanken, A.; Felner, I. *J. Appl. Phys.* 2001, *89*, 6324-6328.
[115] Rice, G. W.; Fiato, R. A.; Soled, S. L., 1987.
[116] Wu, S. H.; Zheng, X. C.; Wang, S. R.; Han, D. Z.; Huang, W. P.; Zhang, S. M. *Catal. Lett.* 2004, *97*, 17-23.
[117] Yu, L.; Zhang, X.; Du, Z.; Wang, D.; Wang, S.; Wu, S. *J. Natural Gas Chem.* 2007, *16*, 46-52.
[118] Yubero, F.; Gonzalez-Elipe, A. R.; Tougaard, S. *Surf. Sci.* 2000, *457*, 24-36.
[119] Yubero, F.; Oca¤a, M.; Justo, A.; Contreras, L.; Gonzalez-Elipe, A. R. *J. Vacuum Sci. Technol. A: Vacuum, Surfaces, Films* 2000, *18*, 2244-2248.
[120] Hansen, T. W.; Wagner, J. B.; Hansen, P. L.; Dahl, S.; Topsoe, H.; Jacobsen, C. H. *Science* 2001, *294*, 1508-1510.
[121] Yang, Y.; Xiang, H. W.; Xu, Y. Y.; Bai, L.; Li, Y. W. *Appl. Catal. A: Gen.* 2004, *266*, 181-194.
[122] Arakawa, H.; Bell, A. T. *Ind. Eng. Chem. Proc. Des. Dev.* 1983, *22*, 97-103.
[123] Raje, A. P.; O'Brien, R. J.; Davis, B. H. *J. Catal.* 1998, *180*, 36-43.
[124] Lohitharn, N.; Goodwin, J. G., Jr. *J. Catal.* 2008, *257*, 142-151.
[125] Hutchings, G. J.; Boeyens, J. C. A. *J. Catal.* 1986, *100*, 507-511.
[126] Bukur, D. B.; Koranne, M.; Lang, X.; Rao, K. R. P. M.; Huffman, G. P. *Appl. Catal. A: Gen.* 1995, *126*, 85-113.

[127] O'Brien, R. J.; Xu, L.; Spicer, R. L.; Davis, B. H. *Energy Fuels* 1996, *10*, 921-926.
[128] Bukur, D. B.; Lang, X. S.; Ding, Y. J. *Appl. Catal. A: Gen.* 1999, *186*, 255-275.
[129] Kölbel, H.; Ralek, M. *Catal. Rev. Sci. Eng.* 1980, *21*, 225-274.
[130] Hofer, L. J. E.; Emmet, P. H. in: (Ed.), Crystallite phases and their relationship to Fischer-Tropsch catalysis, 1956, pp. 407.
[131] Schulz, H.; Schaub, G.; Claeys, M.; Riedel, T. *Appl. Catal. A: Gen.* 1999, *186*, 215-227.
[132] Shroff, M. D.; Kalakkad, D. S.; Coulter, K. E.; Kholer, S.; Harrington, M. S.; Jackson, N. B.; Sault, A. G.; Abhaya, K. D. *J. Catal.* 1995, *156*, 185-207.
[133] Blanchard, F.; Reymond, J. P.; Pommier, B.; Teichner, S. J. *J. Molec. Catal.* 1982, *17*, 171-181.
[134] Dictor, R. A.; Bell, A. T. *J. Catal.* 1986, *97*, 121-136.
[135] Kuivila, C. S.; Stair, P. C.; Butt, J. B. *J. Catal.* 1989, *118*, 299-311.
[136] Amelse, J. A.; Butt, J. B.; Scwartz, L. H. *J. Phys. Chem.* 1978, *82*, 558-563.
[137] Riedel, T.; Schulz, H.; Schaub, G.; Jun, K. W.; Hwang, J. S.; Lee, K. W. *Topics Catal.* 2003, *26*, 41-54.
[138] Eliason, S. A.; Bartholomew, C. H. *Stud. Surf. Sci. Catal.* 1997, *111*, 517-526.
[139] Xu, J.; Bartholomew, C. H. *J. Phys. Chem. B* 2005, *109*, 2392-2403.
[140] Herranz, T.; Rojas, S.; Pérez-Alonso, F. J.; Ojeda, M.; Terreros, P.; Fierro, J. L. G. *J. Catal.* 2006, *243*, 199-211.
[141] O'Brien, R. J.; Xu, L.; Milburn, D. R.; Li, Y. X.; Klabunde, K. J.; Davis, B. H. *Topics Catal.* 1995, *2*, 1-15.
[142] Lohitharn, N.; Goodwin, J. G., Jr. *J. Catal.* 2008, *260*, 7-16.
[143] Pérez-Alonso, F. J.; Herranz, T.; Ojeda, M.; López Granados, M.; Terreros, P.; Fierro, J. L. G.; Gracia, M.; Gancedo, J. R. *Green Chem.* 2007, *9*, 663-670.
[144] Malin, C. B. *Oil Gas J.* 1998, *96*, 33-35.
[145] Eliasson, B.; Riemer, P., *Greenhouse gas control technologies*; Elsevier Science Ltd.: Oxford, 1999.
[146] Hilaire, S.; Wang, X.; Luo, T.; Gorte, R. J.; Wagner, J. *Appl. Catal. A: Gen.* 2001, *215*, 271-278.
[147] Krishnamoorthy, S.; Li, A.; Iglesia, E. *Catal. Lett.* 2002, *80*, 77-86.
[148] Riedel, T.; Schaub, G.; Jun, K. W.; Lee, K. W. *Ind. Eng. Chem. Res.* 2001, *40*, 1355-1363.

[149] Lee, J. F.; Chern, W. S.; Lee, M. D. *Can. J. Chem. Eng.* 1992, *70*, 506-511.

[150] Lee, M. D.; Lee, J. F.; Chang, C. S.; Dong, T. Y. *Appl. Catal.* 1991, *72*, 267-281.

[151] Pérez-Alonso, F. J.; Ojeda, M.; Herranz, T.; Rojas, S.; González-Carballo, J. M.; Terreros, P.; Fierro, J. L. G. *Catal. Commun.* 2008, *9*, 1945-1948.

[152] Hildebrandt, D.; Glasser, D.; Hausberger, B.; Patel, B.; Glasser, B. J. *Science* 2009, *323*, 1680-1681.

[153] Hildebrandt, D.; Glasser, D.; Patel, B.; Hausberger, B. P.; University of the Witwatersrand, J., Ed.: South Africa, 2007.

[154] Boerrigter, H.; Calis, H. P.; Slort, D. J.; Bodenstaff, H. *http://ecn.nl/docs/library/report/2004/rx04041.pdf* 2004,

[155] Olofsson, I.; Nordin, A.; Soderlind, U. *http://biofuelregion.se/dokument/5_95.pdf* 2005,

INDEX

C

D

E

F

G

H

I

J

K

L

M

N

S

T

U

V

W

X

Y

Z